TIMOTHY MORTON

Being Ecological

A PELICAN BOOK

PELICAN BOOKS

UK | USA | Canada | Ireland | Australia
India | New Zealand | South Africa

Penguin Books is part of the Penguin Random House group of companies whose addresses can be found at global.penguinrandomhouse.com.

First published 2018
004

Book design by Matthew Young
Set in 10/14.664 pt FreightText Pro
Typeset by Jouve (UK), Milton Keynes
Printed in Great Britain by Clays Ltd, Elcograf S.p.A.

A CIP catalogue record for this book is available from the British Library

ISBN: 978-0-241-27423-1

For Lindsay Bloxam and Paul Johnson

Contents

‘Grass is hard and lumpy and damp,
and full of dreadful black insects.’

OSCAR WILDE

ACKNOWLEDGEMENTS

Thank you so much to my editors, Thomas Penn and Ananda Pellerin, for their tireless help. In particular, thanks to Ananda for her incredibly detailed and insistently kind editing. Writing in the world of scholarship can be a very isolating affair, so I'm very grateful to have been seen.

Thanks to everyone I gave lectures to or who participated in seminars or workshops or dialogues in the last year. Those occasions are my lab, and I simply couldn't think without them. Thanks to my Dean of Humanities at Rice University, Nicholas Shumway, whose support of me has been one of the miracles of my life. Thanks to my fantastic research assistants, Kevin MacDonnell and Randi Mihajlovic, whose invaluable work enriched this project so much. And thanks again to Kevin for assisting me in every respect; it's not an exaggeration that I just couldn't do my job without his unstinting and incredibly generous support. Thanks to everyone who helped in all kinds of ways, and especially Heitham Al-Sayed, Ian Bogost, Paul Burch, Federico Campagna, Olafur Eliasson, Björk Gudmundsdottir, Sofie Grettve, Lizzy Grindey, Graham Harman, Douglas Kahn, Jeffrey Kripal, Ingrid Luquet-Gad, Edouard Isar, the Morton family (Garth, Jasmine, Charles and Steve), Yoko Ono, Sunny Ozell, Andrea

Pagnes, Sabrina Scott, Priscilla Elora Sharuk, Emilija Škarnulytė, Verena Stenke, Marita Tatari and Cary Wolfe.

Special thanks to Yoko Ono for permission to reproduce part of her work *THIS IS NOT HERE* (p. 116).

This book is dedicated to my cousin, the artist Lindsay Bloxam, and her partner, the artist Paul Johnson. Their genius is to suspend the heavy hand of judgement and open up the world to curiosity, wonder and lightness. We could do with a bit here in the realm of ecological thought.

INTRODUCTION

Not Another Information Dump

Don't care about ecology? You might think you don't, but you might all the same. Don't read ecology books? This book is for you.

It's understandable: ecology books can be confusing information dumps that are out of date by the time they drop on you. Slapping you upside the head to make you feel bad. Shaking your lapels while yelling disturbing facts. Handwringing in agony about 'What are we going to do?' Horseshoe-in-a-boxing-glove propaganda. This book has none of that. *Being Ecological* doesn't preach to the eco-choir. It's for you: maybe you're in the choir but only sometimes, or maybe you have no idea what choirs are, or maybe you don't care at all. Rest assured this book is not going to preach at you. It also contains no ecological facts, no shocking revelations about our world, no ethical or political advice, and no grand tour of ecological thinking. This is a pretty useless ecology book, in fact. But why write something so 'useless' in such urgent times? Have I never heard of global warming? Why are you even reading this? Well, the truth is you might already be ecological, you just didn't know it. How, you might ask? Let's begin and find out.

What This Book is About

In this Introduction, I'll set out the general approach of the book. In the first chapter, I will finger-paint a way of feeling ourselves around the age we live in, which is one of mass extinction caused by global warming. In the second chapter, we'll get on with considering the object of ecological awareness and ecological thinking: the biosphere and its interconnections. In the third chapter, we'll look at what sorts of actions count as ecological. And in the fourth, we'll be exploring a number of current styles of being ecological.

Along the way I'm going to make you familiar with my style of doing philosophy. If that style were a movie directed by me, its producer would be Graham Harman's object-oriented ontology (more on that soon), and its executive producers would be the philosophers Immanuel Kant and Martin Heidegger.

For now, in this Introduction, I'm going to show how this is not an ordinary book about ecology, because it's trying very hard to avoid a seductive rhetorical mode: the guilt-inducing sermon. How? Let's begin with the fact that this book is largely free of facts. I just thought I'd put that in up front myself, before the critics do it.

When you write a book about ecology, whether or not you are a scientist talking about ecological issues, you seem to have to put a lot of facts in. It feels like a requirement of the genre – a genre being like a kind of horizon, a horizon of expectation. We expect tragedies to make us feel certain emotions (Aristotle thought these were fear and pity), and comedies are supposed to make you smile. There is a genre of the kind of writing you find in your passport. And there is definitely a genre of ecological speech – several genres in fact.

Big Other is Watching You

A genre is a sort of world or *possibility space*. You can make certain moves within that space, and as long as you stay in the space, you are performing something in that generic mode. For example, you probably have a certain way of being at a party, and this might be different from your way of being at a company meeting. You might have a certain way of reading the news, and you definitely have certain ways of following (or ignoring) the latest clothing fashions.

Genres are slippery animals. They have to do with what some philosophy calls *the Other* – and when you try to point directly at the other, it (or she or he, or they) disappears. The other – my idea of your idea of her idea of their idea of his idea of my idea of their idea . . . If you've ever been in a band you will know what a perilous concept this is. If you write music tailored to what you think people want in the record store, you might end up paralysed by indecision. This is because the realm of the other is like a network or web of assumptions, prejudices, and preformatted concepts.

Now there are preformatted concepts that are obvious to all of us, or at least they can easily become so. If you want to know the kind of ravioli they make in Florence, then you're going to be able to look that up. 'Florentine ravioli mode' is something you can find out about – indeed, nowadays you can just Google it. *To Google* has at least one meaning to do with this idea of genre. When we Google something, we are often trying to see what the 'other' thinks about it. Google is like the other, some kind of tangled spider web of expectations lurking just out of the corner of our eye, or just on the other side of all those links we don't have time to click. We *never* have enough time to click all the links (as Google gets bigger, this becomes more obvious). Another way of saying this is that this weird thing, the other, is somehow *structural*: it doesn't matter how you sidle up to it, you will never be able to grasp it directly. Its job seems to be to disappear whenever you look directly at it, but to feel like it's surrounding you when you don't – sometimes this feeling can be pretty creepy.

Who are We?

I'm going to be saying *we* a lot in this book. It's not fashionable to say *we* in my line of work (humanistic scholarship). It's fashionable to be very explicit about how different people are, and it's considered to be passing over or even erasing those significant differences to say *we*. In addition, pronouns are complicated things in an ecological age: how many beings does *we* gather together and are they all human? I'm going to be using *we* as someone thoroughly informed by the politics of difference, and by the identity politics that distorts it. I'm going to be using *we* in part to highlight how the beings responsible for global warming are not seahorses: they are humans, beings like me. It's about time we figured out how to talk about the human species, while at the same time not acting as if the last few decades of thought and politics had never happened. We surely can't go back to imagining some vanilla essence of 'Man' underneath our differences. But if we don't figure out how to say *we*, someone else will. And as the Romantic poet William Blake said, 'I must create my own system or be enslaved by another man's.'

Facing Facts

We all know that ecological writing – especially the sort that delivers scientific information, maybe the kind you often find in a newspaper, but definitely the kind you also find in books with titles like this one – needs lots of facts. Lots of *data*. You would be right to think that this data is usually delivered in a certain mode, once you stop to think about it – but no one is stopping to think about it very much. 'Ecological information delivery mode' has a certain flavour, a certain style – it happens in a certain *possibility space*. One of my jobs as a Humanities scholar is to try to feel out these possibility spaces, especially if/when we're not very aware of them. Possibility spaces that aren't very obvious to us can exert all kinds of control over us, and we may not want these kinds of control – or at any rate, it might be nice to get a sense of what the coordinates are. Just think about the long history of sexism or racism: they have affected our behaviour in all kinds of ways we may not be aware of – and it has taken a lot of time and effort from a lot of different people to make obvious the types of patterns of thought, assumptions and behaviour that underlie prejudice and even make people think it's OK.

What are the laws of gravity in the possibility space? Which way is up, which way is down? What counts as wrong, what counts as right? How far can you venture within the space before you cross over into another space? For example, how far can you distort ecological information mode before it turns into something else? That might actually be a good way of finding out what a possibility space is, just like it's a good idea to find out what a metal is by heating it, freezing it, firing pulses of energy at it, putting it in a magnetic field and so on – the old image of biting into a gold coin comes to mind. It's the same with art. You can find out what a play is like by imagining how far you could distort it before it really does become something quite different. How many crazy costumes can you get away with – if you staged Shakespeare's play *Hamlet* on Jupiter using people dressed as hamsters, would we still recognize it as *Hamlet*?

Perhaps my intentions might be more obvious if I put it this way: this book is free of *factoids*. A factoid is a fact that we know something about – we know that it has been coloured or flavoured a certain way, that it's supposed to look and quack like a fact. Perhaps it's even true, at least from one or more points of view. But still, it has a strange quality. It seems to be shouting at us – *Look. I'm a fact. You can't ignore me. I dropped out of the sky on your head.* That's interesting – a fact that was designed to look like it dropped out of the sky. Factoids are designed to look like what we think facts should be – we think they should look like they are not designed. When people use factoids, we feel like we are being manipulated by little bits of truth that have been broken off some larger, truer edifice, as if they were small chunks of cake.

Consider for example the *factoid* that 'there is a gene for' some trait. Most people take this to mean that a part of your DNA code will cause you to have this trait. But when you study evolution and genetics, you will find out the *fact* that *there are no 'genes for' anything*. The *fact* is that traits emerge through complex reactions between DNA expressing itself and the environment in which the DNA is doing the expressing. Just because you have some DNA that is associated with a certain cancer, it doesn't mean you will get it. But we go around repeating the factoid that 'there is a gene for this or that cancer'.

How We Talk to Ourselves about Ecology

Ecological information delivery mode in the media seems most often to consist of what we could call an *information dump*. At least one factoid – and often a whole plateful – seems to be falling on to our heads. And this falling has an authoritative quality: the delivery mode seems to be saying *Don't question this*, or even *You should feel very bad if you question this*. In particular, 'global warming information mode' seems to be about dumping massive platefuls of facts on to us. Why? This is another way of saying *What are the moves we can make in the possibility space of global warming information mode?* Which is a rather complex way of saying *What is the genre of global warming information mode?* Which way is up? How are we supposed to feel? What kind of information delivery would destroy this mode? And so on.

Our not having a ready answer for this question, unless

we are global warming deniers, should make us pause. Deniers are quite clear: this mode is trying to convince me of something I don't want to believe. I am having a belief forced down my throat. Why don't we all feel like that? And if we feel ecologically righteous, we shun people who think they are being dumped on to make them feel something – crude guilt leading to crude belief, maybe. This is not a war of beliefs – this is the truth. Damn it, Mr Denier, why can't you see that?

Despite what factoids would have us believe, no fact just plops out of the sky. There is a whole environment in which the fact can appear – otherwise you can't see it at all. Consider something you might not regularly say if you grew up in the West: *My ancestral spirits are unhappy that I'm writing this book*. In what world does this statement make sense? What do you need to know, what do you need to expect? What counts as right and wrong in this world? We need all kinds of assumptions about what reality is, about what counts as real, what counts as existing, what counts as correct and incorrect. Thinking about these kinds of assumptions can take different forms, in philosophy one is called *ontology*, another is called *epistemology*. Ontology is the study of how things exist. Epistemology is the study of how we know things.

In addition to the idea that facts are meaningful within certain contexts of interpretation, there are questions you can answer quite easily if you study art, music or literature. These are questions such as *How does the mode want you to read this information? How do you look like you received it 'right'*? You don't look at a Renaissance perspective painting from the side. You have to stand pretty much right in front of the vanishing point, at a certain distance – then the 3D illusion makes sense. The picture positions you in a certain way, the poem asks to be read a certain way – just like a Coca-Cola bottle 'wants' you to hold it a certain way, a hammer seems to fit your hand just so when you handle it . . . A whole lot of what is sometimes called ideology theory is about how you are coerced into handling a poem, a painting, a political speech, a concept in a certain way.

All kinds of ontology and epistemology (and ideology) are implied by ecological information dump mode, but we rarely pause to figure out what they are. We are too keen on dumping, or being dumped on. Why? Why don't we even seem to want to pause and figure it out? Are we scared we might find something? What are we scared we might find? Why do we wring our hands and go *Why don't these deniers get it?* or *Why doesn't my neighbour care about all this as much as I do?* Ecological information dump mode is a symptom of something much bigger than feelings about stuff you read in the newspaper.

One way to zoom out and ask these sorts of question again would be to say something like *How are we living ecological data? Do we like it? If not, what do we want to do about that?* This book, *Being Ecological*, is about how to *live* ecological knowledge. It seems to be not enough just to know stuff. In fact, it seems like 'just knowing stuff' is never just knowing stuff, according to what I've been trying to argue. 'Just knowing stuff' is a way of living things too. And knowing that there is a way of living things implies there could be other ways too. If you have tragedy, you can imagine something like comedy. If you live in New York, you can imagine living in not-New York.

There seem to be plenty of ways of living ecological knowledge. Just think about being a hippy, something with which I am vaguely familiar. Being a hippy is a whole way of life, a whole style. But is being a hippy compulsory as a way to live ecological information? Think about the internet. Before a huge number of people had access to it, there were two or three ways of living with the internet. For instance, there was the amused, playful, experimental, anarchic or libertarian slacker mode in which the internet was supposed to make us feel like our identities were malleable or liquid. Then something strange happened. Loads more people got the internet, and a whole lot of the internet became a really coercive, authoritarian space where you had to have one of about three acceptable opinions or risk being attacked by a mob of judgemental twitterers like the flock descending on the gas station in Alfred Hitchcock's film *The Birds*. I'm not going to go into why and how this happened, but you get the point.

Being Ecological is starting by peering under the hood of the ways in which we talk to ourselves about ecology. I think the main way – just dumping data on ourselves – is actually *inhibiting* a more genuine way of handling ecological knowledge. There are better ways of living all of this than we have now, and we don't even *know* that we are living it right now. We are like people caught in a habitual pattern, going along repeating the same thing, without even realizing it. It's like we find ourselves at the sink, compulsively washing our hands over and over again – but we have no idea how we got there.

Facts go out of date all the time, especially ecological facts, and especially, out of those, global warming facts, which are notoriously multidimensional and scaled to all kinds of temporalities and all kinds of scenarios. Dumping information on ourselves every day or every week can be really confusing and arduous. Imagine it from another angle. Imagine that we are *dreaming*. What kind of dream would it be where the characters and plot vary, sometimes significantly, but the overall impact – where the dream leaves us, its basic colour or tone or point of view (or what have you) – remains the same? There is definitely an analogy from the world of dreaming: these are the trauma dreams of sufferers of post-traumatic stress disorder (PTSD).

Ecological PTSD

In PTSD dreams, you imagine yourself re-experiencing your trauma, and the dreams have a nasty habit of recurring. The founder of psychoanalysis, Sigmund Freud, wondered why this was the case – how come we dream about things that seem to be harmful to us, in a dreaming mode that is also apparently harmful in some way? – it shocks us, it wakes us up crying or sweating, we can't shake it off as we go about our daily business, and so on. Freud argued that there must be some kind of *pleasure* in such a process, otherwise we wouldn't be doing it to ourselves.[1] There must be some aspect of dumping trauma data on ourselves in the world of dreams that is *pleasurable*. And if my analogy holds, this means that information dump mode is in some way enjoyable, as confusing and oppressive as it so often seems.

The PTSD sufferer, Freud argued, is simply trying to install herself, through her dreams, at a point in time *before the trauma happened*. Why? Because there is some safety or security in being able to *anticipate*. Anticipatory fear is far less intense than the fear you experience when finding yourself, all of a sudden, in the middle of a trauma – Freud calls that kind of fear *fright*. If you think about it, traumas by definition are things that you find yourself in the middle of – you can't sneak up on them from the side or from behind, and that's why they're traumatic. You just suddenly find yourself in a car crash, for instance. If you had been able to anticipate, you might have been able to swerve out of the way.

PTSD dreams are trying to create a blister of anticipatory fear (Freud calls it 'anxiety' – which is a bit confusing in terms of this book, so I won't say it again) that will surround the raw trauma of fright. By analogy, then, information dump mode is a way for us to try to install ourselves at a fictional point in time *before global warming happened*. We are trying to anticipate something inside which we already find ourselves.

Doing Something

The explicit content of the data seems so urgent: it's as if it is screaming, 'Look, can't you see? Wake up! Do something!' But the *implicit content* of the mode in which we send and receive this data contradicts this urgency in a stark way: 'Something is coming but it's not here yet. Wait – look around, anticipate.' Can you see how the message is two-faced? One face is shocking, urgent; the other face is an anti-shock blister. What does this mean? It means that no amount of refining the data or data dump mode will ultimately work. It's impossible to get a PTSD dream to line up with the fright it's trying to transmute. In exactly the same way, ecological information dump mode (it doesn't just affect global warming) is, and I need to say this in the most contrastive way possible, *exactly the opposite* of what we need in order to comprehend where we are and why – to *start to live the data*. Right now it's as if we are waiting for just the right kind of data, then we can start living in accord with it. But this data will never arrive, because its delivery mode is designed to prevent the appropriate reaction – we find ourselves in the midst of horribly confusing, traumatic events such as global warming and mass extinction, and we don't have much of an idea of how to *live* that.

Isn't this PTSD mode the real reason why it seems so difficult to do something, anything? Almost every environmental studies conference I go to ends with a round-table during which someone will pipe up, 'But what are we supposed to be *doing*?' As if worrying for days about something isn't actually already a form of 'doing'. The 'What are we going to do?' is a symptom of finding ourselves in a frightful situation, frightful in Freud's technical sense of realizing we are going through a trauma. As with all traumas, we didn't realize how horrible it was until we were some of the way into the experience. What we don't want to know is this 'It's already happening' quality of the ecological emergency. The question at the end of the round-table wants to *see ahead* and anticipate and know what to do, in advance. That's what we can't do. Because we have been driving the wrong way, looking in the wrong direction – that's exactly *why* all this happened.

Ecological facts are, at present, very often facts about the *unintended* consequences of human actions. Exactly: the vast majority of us had no idea what we were doing, on some level. It's like the *noir* movie where the lead character discovers that all along she was working for a hostile secret agency.

So I have a lot of sympathy for the 'What are we going to do?' sort of question. And this is precisely why I refuse to give it a straight answer. What this type of question is asking, and the way the question is asking it, has to do with needing to control all aspects of the current ecological crisis. And we can't. That would require being able to reverse time and return at least to 10, 000 BCE, before humans set the agricultural logistics in motion that eventually gave rise to the Industrial Revolution, carbon emissions and therefore to global warming and mass extinction.

But there is a quite benign explanation for all this, in a sense. It's *never* the case that you think first, then act. You can't see everything all at once. You just sort of muddle around, and then you get some kind of snapshot of what's going on, with more or less accurate hindsight. Foreseeing and planning are strangely overrated, as neurology is now telling us, and as phenomenology has been telling us. It has to do with how we overrate the idea of free will. Our agricultural-based religions tell us that we have a soul that is somewhere inside yet beyond our body, and that this soul guides the body around, like a charioteer steering the horses (this is how the Greek philosopher Plato puts it in *Phaedrus*). But this idea has its origin in the very dynamic we have identified as the problem. We've been thinking that we are on top of things, outside of things or beyond things, able to look down and decide exactly what to do, in all sorts of ways for about 12,000 years.

Maybe ecological facts require that we *don't* immediately 'know' exactly what to do.

Yet there is a paradox: it's so very clear that 'what to do' is drastically to limit or eliminate carbon emissions. We know *exactly* what to do. Why aren't we doing it? There are great ways to let yourself off the hook here. For example, you can argue that neoliberal capitalism is so oppressive and all-pervasive that it would require a major global revolution to dismantle the structures that are polluting the biosphere with carbon emissions: the big corporations. So there should be a gigantic social revolution first, then once we have the right way of relating with one another, we can get down to the business of curbing our emissions. Isn't this weirdly the same as the argument India made at the climate negotiations in Copenhagen in 2009? India argued that it couldn't limit emissions, because first it needed to go through exactly the same kind of 'development' as the West. Once it had achieved the right kind of society, it could think about curbing its harmful ways.

Assuming that this strategy actually works, by the time you have achieved what you wanted, Earth will have melted anyway.

Things versus Thing-Data

The 'What are we going to do?' question is weird: there is a very accurate description of what to do, yet it will never, ever feel as if we are doing it exactly right, even if we try. Here is the paradox: we know what to do *and* we won't be able to get high up enough above the world to see exactly what that looks like. And it's very strange, because these two facts go together: we have accurate data and accurate solutions, yet – *and* – this goes along with being unable to see the wood for the trees. There always appears to be too many trees.

By the way, the problem is much more 'interesting' (aka worse) than I've just described. This is because *any action at all* will suffer from this paradox. Say for example you 'know what to do' and that involves individuals or small groups limiting their emissions, rather than dismantling global capitalism or sidestepping the polluting aspects of modern modes of production. You will never be able to check in advance as to whether your actions are having the desired effect, and in particular you know that Earth is so large that your small action won't count for much, if anything. In fact, your own personal emissions are probably statistically meaningless. But billions of them are exactly what is causing global warming. This is what the data is telling you. Yet doing nothing at all is exactly the problem, so just feeling smug and powerless won't work either.

'What are we going to do?' wants to be relieved of something. What? It wants to be relieved of the burden of anxiety and uncertainty. But data in general is all about anxiety and uncertainty, let alone global warming data. This is because data is statistical. You will never be able to prove that *x* definitely causes *y*. The best you can do is say that it's 99 per cent likely that *x* is causing *y*. So for example the patterns in the cloud chambers at the Large Hadron Collider, the particle accelerator in Geneva (CERN), that are evidence for the

Higgs boson might not fully prove the existence of this elementary particle: it's just that this 'might not' is confined to a tiny fraction of a decimal part of 1 per cent of the probability range. If you think about it, this is much better than just asserting stuff, because it means noticing things that are real, and it also means you don't have to back up your assertion with the threat of some kind of violence. There *is* a Higgs boson, not because the Pope is forcing you to believe in it but because it's incredibly unlikely that there isn't one, based on the patterns physicists are seeing in the data. That's what scientists do: they look for patterns in data. Looking at patterns: it's a lot more like appreciating art than you might think – more on this later.

Truthiness

Data simply means *what is given*. It's the plural form of the supine of the Latin *dare, to give*: aspects of things that are *given* to us when we observe them. If we have a pair of scales, we can collect data about the weight of an apple. If we have a particle accelerator, we can collect data about the protons in the apple. In truth, data isn't really the same as facts, let alone interpretations of facts. In order to have a fact, you need two things: data, and an *interpretation* of that data. This sounds counterintuitive, because part of our common talk about science thinks in a very old-fashioned way about facts. Common talk imagines facts to be something like barcodes that you can read off of a thing: they are self-evident. But a scientific fact isn't self-evident. That's precisely why you have to do an experiment, collect data and interpret that data.

Notice that neither data nor interpretations are the actual *things* about which we are gathering data and interpreting. A *factoid* is a (usually quite small) chunk of data that has been interpreted so as to *appear* truthful. It is 'truthy', to use the helpful vocabulary of the American comedian Stephen Colbert and his parodic word, 'truthiness'. It has a ring of truth, or as some scientists now say, it is 'truth-like'. A factoid is truthy because it is in accord with what we think facts are. And because of *scientism*, the common belief that science tells us something about the world in the same way that a religion might do, we think that facts are totally simple and straight: they come out of things themselves. Scientism is the worship of factoids. Factoids imply a certain attitude, and that attitude is that things themselves have a sort of barcode on them that tells us immediately – that is, without the mediation of humans interpreting them – what they are. What appears truthy to us is what cuts out the middleman, offers up straight data. But data isn't facts – yet. And ecological data is so complex, and is about such complex phenomena, that it's difficult to make that data into facts, let alone to start living those facts, rather than repeating truthy factoids, which are the contents of the PTSD dream we keep indulging in. There is an exasperated 'Can't you *see*?!' about the way this truthiness works. But 'seeing' is precisely what we don't seem to do with this data.

So I'm afraid that the world of science is actually shifty and uncertain. And any attempt to achieve total certainty is an attempt not to live in a scientific age. Data dump mode, even if we accept global warming is real, will never give us the satisfaction we think we want. We spew it and listen to it as if it could, and that's the problem. We are stuck in the initial stages of going through a trauma – one that is still happening, mind you, one whose painfulness is obvious if you care at all. It's like trying to have a PTSD dream *while having a trauma*, as if you could go to sleep and dream that you were anticipating the approaching car at the exact moment at which you were crashing. Putting it this way, can you see how the mode in which we most often get caught in news reports, press conferences, dinner conversations and books such as this just isn't helping at all?

Denial of planetary syndromes such as global warming bogs us down in factoids. We waste a lot of time worrying about or arguing with these factoids, which have nothing to do with data and interpretations of data. When we get into this mode – either of being deniers or arguing with deniers – we are barking up the wrong tree. Truthiness is in a way a kind of reaction, like a blister, to the real problem, namely that we live in a modern scientific age characterized by a radical gap between data and things. No one access mode can exhaust all the qualities and characteristics of a thing. Therefore things are open, they withdraw from total access. With your thought you can't encapsulate everything that an apple is, because you forgot to taste it. But biting into an apple won't capture everything an apple is either, because you forgot to tunnel into it like a worm. And so with tunnelling too. What you have, in each case, is not the apple in itself, but apple data: you have an apple thought, you have an apple bite, you have an apple tunnel. A diagram of every possible access to the apple throughout all of time and space – assuming it could be made (which it couldn't) – would miss the kind of apple that a less complete diagram would capture. And in both cases you wouldn't have an apple, you would have an apple diagram. But for sure there is apple data: apples are green, round, juicy, sweet, crunchy, packed with Vitamin C; they make an appearance in Genesis as the most unfortunate snack in human history, they sit on boys' heads waiting for arrows to shoot them in stories . . . None of these things are the apple as such. There is a radical gap between the apple and how it appears, its data, such that no matter how much you study the apple, you won't be able to locate the gap by pointing to it: it's a *transcendental* gap.

The transcendental gap between things and thing-data becomes quite clear when we study what I like to call *hyperobjects*: things that are huge and, as they say, 'distributed' in time and space – that take place over many decades or centuries (or indeed millennia), and that happen all over Earth – like global warming. Such things are impossible to point to directly all at once. Such things (evolution, biosphere, climate, for example) give us a clue about how things are – everything, according to our modern way of looking at them. Everything: a spoon, a small plate of scrambled eggs, a parked car, a soccer pitch, a woolly hat. None of these things can be pointed at directly. When you feel your woolly hat, what you are feeling is woolliness – you are receiving hat data, not the actual hat. When you put it on your head, you are *using* or *accessing* the hat in a certain way – but you aren't completely accessing it. As it begins to warm your head and you get on with your morning walk in the cold air, your hat sort of disappears – you are busy with getting from A to B, and now you're nice and warm, so the hat is doing its job and you can forget about it. This quality of things – which sort of disappear when they are functioning nicely in your world – should give you a clue as to what they actually are. What things actually are is sharply different from thing-data. When you look at your hat or photograph your hat, you have a hat vision or a hat photo, not the actual hat.

A hat factoid is pretending to be about an actual hat. But a hat factoid is a certain interpretation of hat data, pretending not to be an interpretation. This way of being true is actually seriously out of date: more than two hundred years out of date at present. David Hume, the renowned Scottish philosopher of the later 1700s, argued that you just can't peer directly under the lid of data to get at what things actually are. And his immediate successor in the later eighteenth century, the philosopher Immanuel Kant, explained why: it's because of this radical gap I've started to talk about, the gap between things and data. Ecological things are very complex, involve a lot of moving parts, are widely distributed across Earth and across time, and so on. So peering under ecological thing-data is obviously impossible – we get confused when we try.

Including Our Perspective in the Picture

Global warming denial is actually a displaced sort of modernity denial. There is something we don't want to know very clearly about our modern age, and that something is what Hume and Kant were talking about. Data is shifty, data isn't things, and data is all you've got. I sometimes wonder whether Hume was reincarnated as Roger Waters, the bass guitarist and lyricist of Pink Floyd. On their album *The Dark Side of the Moon*, on the song 'Breathe', they sing something that Hume could easily have written himself: 'All you touch and all you see / Is all your life will ever be.'[2] That's exactly it. You don't get to handle things directly, without hands and eyes – and by extension without experimental apparatus, thermometers, laboratories and ideas about what scientific facts are. Funnily enough, living in a scientific age means that you realize more and more that you are shrink-wrapped in your experience.

'Natural' Means 'Habitual'

The Romantic poets, who lived around the time of Hume and Kant, got a handle on this quickly. They realized that when you get really up close to things, they start to 'dissolve'. This is another way of saying that when you let go of a normalized reference frame, the strangeness of things, the way you can't access them directly, becomes very obvious. Say, for example, you are examining a rock face with a geologist's hammer and a magnifying glass. You are so much closer to that rock than someone who is looking at it in a picture postcard. The picture postcard looker is pretty sure of what she is seeing. Picture postcards are descendants of what came before Romanticism in art, namely the picturesque. In the picturesque, the world is designed to look like a picture – like it's already been interpreted and packaged by a human. You can easily see what's what: there's a mountain over there, a lake, maybe there's a tree in the foreground. Funnily enough, the classic picturesque image, which I have just described, is on average everyone's favourite image – everyone on planet Earth, and maybe its ubiquity is why many people also find this image kitsch or obvious. And funnily enough, this is pretty much what humans saw in the savannah millions of years ago. Having a body of water nearby and some shade (those trees), encircled safely by mountains where you know there is water descending to feed the lake (for instance), is pretty handy if you're some kind of ancient human. The picturesque is keyed to a fundamental human-centred way of looking at things: it is *anthropocentric*.

But the view of the mountain from close up, that's a whole different matter. Say you are a Romantic poet or a scientist and you decide to take off and walk into that picture, into that 'landscape' – which means a *picture* of a landscape. The picture quality evaporates. Now you are up close and personal with the rock. It stops being a nice background to your Paleolithic projects as an ancient human. It starts to become quite strange: you see all kinds of crystals, all kinds of curves and shapes that don't have much relevance to your regular world. You may begin to see fossils – other lifeforms have been using this rock in a different way from you. Or perhaps you notice that a bird has made a nest in a crevice. You start to realize that this isn't just your very own world.

It's like having jet lag. When you arrive in a very distant place, you are a little bit freaked out (or a lot freaked out) by the simple fact that this place isn't yours, not yet. In fact, you are so tired and your body clock is so upside down, even the time isn't yours. Time stops being a nice neutral box that you just live in and forget about, waiting for the alarm or the calendar to remind you of what to do and when. Time stops being what it actually isn't – namely, a human interpretation of time. 'Interpretation' doesn't just mean 'mental description'. It means the whole panoply of ways in which you access and use a thing. How you access an apple gives you apple data, remember, not apples in themselves. Even eating the apple gives you apple bites, not the entire apple in all its manifold glory. Think of how we like to talk about 'interpretations' of music. That doesn't mean simply thinking about the music – it also means actually *playing* the music: executing the music. The conductor of the Berlin Philharmonic Orchestra 'interprets' a musical score by waving her arms in the air, causing musicians to 'interpret' the lines of music in certain ways. When you put it that way, it becomes quite obvious. *An execution of a thing is not the thing.*

So, there you are with your geologist's hammer and your special camera, and you have come up against the fact that hammerings and photographings of things aren't those things. Your picturesque world was so consistent that you forgot that this picturesque-ing was also an execution of things like lakes and trees and mountains. You thought you were seeing something directly: you probably call it *nature.* Nature sort of means something you forget about because it's just functioning. We talk about 'human nature' this way. 'It's in my nature, I can't help it.' 'Doing what comes naturally.' And we talk about nonhuman 'nature' this way: that's the whole point of the 'weather conversation' you have with a stranger at a bus stop. You are able to find common ground in something that appears neutral, something that just functions and therefore creates a background for your interaction. But global warming takes that supposed neutrality away from us, like too-eager stage hands removing all the scenery while the play is still in progress.

So your scientific view of things, up close with a hammer and a camera, doesn't mean you're 'seeing' nature; you are still interpreting it with human tools and a human's touch. Thinking in an ecological way means letting go of this idea of nature – it sounds incredible, but that's only because we're so habituated to certain ways of accessing and executing and otherwise 'interpreting' things such as lakes, trees, cows, snow, sunshine and wheat.

The Romantic poets figured out that when you get 'scientific', as I was just describing, when you become open to all kinds of data, not just clichéd stuff, you must also get 'experiential'. You end up writing poems about the *experience* of encountering the rock, and how strange that actually is. You might go a bit further and write a poem about writing a poem about the experience of encountering the rock. This isn't actually unscientific at all. This is how living data works. You realize that you are included in the interpretation, so your art becomes 'reflexive' – it starts to talk about itself. So all this bludgeoning business – all these information dumps – are exactly how not to live scientific data. But they are how we try to override the strangeness of living in a scientific age. They are our reactions to the heaps of information we receive, the things we design and create, the disconnection we feel from nature or ecology, and similarly the panic we experience or the helplessness we feel when we start to think about things like global warming. You can't get to this reflexive mode if you start with a mentality that thinks ecological information is about dumping factoids on people.

Lots of ecological writing, which we often call 'environmentalist', has the same format, roughly speaking, as information dump mode. It's designed to be 'truthy', to put you in touch with something like picturesque Nature – I'm going to start putting that word in capital letters to remind you that this isn't actual trees and bunny rabbits, it's a concept, an interpretation. Funnily enough, twisty, weird, possibly postmodern art is much more up to speed with living in a scientific age than sentimental 'obvious' images of majestic big cats or lush rainforests in one of those glossy photos in a calendar. Living ecological facts is difficult: maybe ecological facts require that we *don't* immediately 'know' exactly what to do. Let's put it more strongly. Maybe they even require that we *shouldn't* immediately know what to do. Add to this the fact of anthropocentrism – for quite some time we have been designing and interpreting and executing things so as to make sure humans are in a top or central position in all the domains of existence (psychic, philosophical, social). Ecological facts are about the unintended consequences of anthropocentrism. So because ecological facts are about us, about how we are and what we do and how we act, they are hard to see from a distance – getting perspective about yourself, interrogating your way of doing and seeing, is one of the hardest things to do – and difficult to swallow, intrinsically.

If you are committed to the reality of what human carbon emissions are doing, don't be so hard on deniers of global warming. You have more in common with them than you might think. Trying to override them with facts presented as factoids is exactly the mode they are also in, which is about hiding from the weirdness of our modern scientific age. You will be fighting fire with fire – or better, cold water with cold water, because factoid speech is trying to pour cold water on the fire of contemporary knowledge, which burns through so many of our assumptions and certainties.

What, then, is ecological reality? I shall be exploring this in the second chapter, where we will consider the most basic ecological fact of all: the fact that lifeforms are interconnected. This seemingly obvious fact is so much stranger than you think.

Why Should *I* Care?

Different cultures have different ways of being a student. Over the years I've noticed this in my travels around the USA, working in four distinct locations (East, Central, West and South). And when I teach seminars in Europe and elsewhere, there are vivid differences too. Students in Paris are quite different from students in Taiwan, who in turn are very different from students in northern California. For example, the difficulty of teaching students in the beautiful high-altitude mountain town of Boulder, Colorado, was that you had to convince them that the poem we were examining was the most psychedelic thing they were ever going to encounter in their lives, because their main extra-curricular activity was getting high on cannabis and going snowboarding. But you had already done this with the previous poem, so you had to keep on upping the ante.

California was quite a shock at first. The basic atmosphere was a kind of nervousness, masked by an affected blasé indifference. It was as if my students were holding an invisible TV remote, and saying silently, 'Amuse us, or we'll change the channel.' Teaching involves working with all kinds of emotional energy, but basically there are about three main flavours, and you relate to them in sequence. They are strawberry, chocolate and vanilla, otherwise known as passion, aggression and ignorance, just like the general Buddhist emotional typology. (There are all kinds of sub-flavours, just like you can have strawberry with a vanilla centre or chocolate with toffee and so on.)

First you want your students to like you – and you want to like your job, so you are working with passion. Then you allow yourself to dislike your job a bit, and you start working with aggression, which means you learn how to let students hate you a bit. You learn how to work with scapegoat energy – the way a group tries to dump its negativity into one person. If this person is a student in a class, he or she becomes the devil's advocate and tries to pick a fight with you in front of everyone, which you learn to deflect and feed back to the class without getting involved.

Finally, you end up working with ignorance or indifference, which is the hardest energy to work with, because the opposite of love isn't hate, it's this vanilla feeling of not caring all that much. It's very tricky, because you can't seem to break into it – that would involve evoking the aggression energy, and your students don't want to go there; or maybe you try to plead with them, using passion, and that makes you feel really vulnerable, and indeed your students will probably ignore your efforts and make you feel frustrated.

This is precisely the trap into which I walked one afternoon. I was teaching something about Romantic art, and I was talking for some reason about pianos, the pianoforte having been invented in the later eighteenth century. I asked something like 'Who knows anything about the history of the piano?' And it happened. The California students are unsure of themselves, yet vocal (and there they sat, holding their invisible remotes). From the right-hand side of the class (yes, reader, I am recounting a vividly remembered trauma), towards the back, higher up, there came a female voice: 'Why should *I* care?'

I felt like I had been slapped.

It had never occurred to me, the good schoolboy, not to care about something that happened in a classroom. The question might as well have been in Martian, so incomprehensible was it at first. I was stunned. I hadn't felt stunned in a classroom for a while, and by then I'd been teaching for about fifteen years. And this was a very new kind of stunned. I hadn't exactly been attacked. At first I didn't know what had happened at all. It was only week two of a ten-week term: this was a bad sign. In the moment, I could think of nothing to say in reply.

That episode haunted me for days. I just couldn't figure it out. It was like having eaten something that's very difficult to digest. But by the end of the week I realized something important. I could apply the same phrase to myself. Why should I, Timothy Morton, care so much about teaching about pianos that it kills me when someone says 'Why should *I* care?' Wouldn't it be better to be a little bit 'care-less', in other words, carefree? And perhaps if you're a bit of a control freak like me, being carefree and open feels a bit like being careless . . . Once upon a time, while I was in Boulder, I had seen a magnificent calligraphy by a Buddhist teacher called Ösel Tendzin in the hallway of my friend Diane's house. With a huge brush, he had drawn two words, disconnected yet connected: CARE LESS. That summed it up. For me, whom Buddha would definitely have pegged as a too-tight person, getting meditation just right always feels like getting it a bit wrong. I now take this feeling of screwing up to be a signal that I'm meditating just right.

As it turned out, the student in question ended up not-caring enough in other situations to lower her grade significantly. But I had learned something valuable.

This book is about caring, so my encounter with that student is highly relevant. Every day, as I've been arguing, we get bombarded by ecological factoids, and ecological issues are truly urgent, and if you think about them too hard, you can become really depressed and end up in the foetal position, or simply curled up in denial like a hedgehog. So I've written this book with a CARE LESS sort of attitude, and I expect you to CARE LESS too. Please don't hit delete on your indifference. Instead, why not study it as we've been doing? You might find that its cloudy realms contain a soft, rubbery ball of numbness. Numbness is a feeling of protecting yourself from a shock. Be very careful with this numbness. Again, don't try to peel open the rubber or stab it with scissors to try to get at what's inside. Instead, try to study it from the outside. Plenty of objects are like that in an obvious way: there's no way, for example, to climb inside a black hole to study it and live, let alone exit to tell others what you found. You have to study phenomena around the black hole, up to and including its event horizon, the point beyond which you simply won't be able to get out and tell the tale.

Object-Oriented Ontology

I adhere to a philosophical view known as *object-oriented ontology* (OOO), which holds that, in many ways, everything is like a black hole: a rubber ball, an emotion, a sentence about an emotion, an idea about a sentence, the sound of that sentence as spoken by a computer, the computer's glass screen, the beach from which the sand that made the glass screen was extracted, ocean waves, salt crystals, whales, jellyfish and coral. You have to study the phenomena these things emit – the philosophical term is *phenomenology* – because you're never going to get at them in themselves. No access mode will work properly: thinking, stabbing with scissors, eating, ignoring, writing a poem about, crawling across (if you are a fly), kicking (if you are a football player), eating (if you are a dog), irradiating (if you are a gamma ray).

OOO was first formulated by an American philosopher, Graham Harman, who was pondering how the philosophy of Martin Heidegger actually works (no matter what Heidegger himself said about it). OOO argues that nothing can be accessed all at once in its entirety.[3] By access is meant any way of grasping a thing: brushing against, thinking about, licking, making a painting of, eating, building a nest on, blowing to bits . . . OOO also argues that thought is not the only access mode, and that thought is by no means the top access mode – indeed, *there is no top access mode*. What these two insights give us is a world in which anthropocentrism is impossible, because thought has been extremely closely correlated with being human for so long, and because human beings have mostly been the only ones allowed to access other things in a meaningful way. OOO offers us a marvellous world of shadows and hidden corners, a world in which things can't ever be completely irradiated by the ultraviolet light of thought, a world in which being a badger, nosing past whatever it is that you, a human being, are looking at thoughtfully, is just as validly accessing that thing as you are.

I think that object-oriented ontology is really useful for an age in which we have come to know much more about ecology. One way is that it doesn't make thinking, in particular human thinking, into a special kind of access mode that truly gets at what a thing is. OOO tries to let go of anthropocentrism, which holds that humans are the centre of meaning and power (and so on). This might be useful in an era during which we need to at least recognize the importance of other lifeforms.

Maybe our indifference – the fact that we don't (want to) care very much (or all the time) about ecological things – is like a unique lifeform, sort of living rent-free in our heads. Maybe we could get much more information about ecology and ecological politics, art, philosophy and culture from studying that cloudy realm containing the rubber ball of numbness, than from trying to crack it open. Maybe we already have everything we need to cope with an ecological age. Maybe the actual problem has been that we keep telling ourselves that we need a totally new way of looking at things because the ecological age is some kind of apocalypse where our familiar world is totally ripped apart. But is this hoping for a new way to see or be really ecological, or is it just a retweet of the agricultural-age monotheism that has got us to this stage in the first place? And if agriculture is in part responsible for global warming and mass extinction (which it is), wouldn't it be better not to use a monotheist reference frame or monotheist language? Wouldn't it be better to stop with the sermonizing, the shaming and the guilt that are part and parcel of the theistic approach to life that arose in the agricultural age?

These are just questions. Please don't care too much about them. As you move through this book, watch any feelings of guilt that come up. After all, guilt is scaled to individuals. But individuals are *in no sense guilty* for global warming. That's right – you can totally let yourself off the hook, because starting the internal combustion engine of your car every day is statistically meaningless when it comes to global warming. The paradox is that when we scale actions like that up to include every car motor start on every day since the internal combustion engine was invented, humans are causing global warming. Big corporations are obviously capable of having this effect. But their employees' effect is, to use the phrase again, statistically meaningless. Several thousand years from now, nothing about you as an individual will matter. But what you did will have huge consequences.[4] This is the paradox of the ecological age. And it is why action to change global warming must be massive and collective.

What is global warming anyway? The correct answer is that it is *mass extinction*. This will be our next topic.

CHAPTER 1

And You May Find Yourself Living in an Age of Mass Extinction

Exactly what is the current state of play, ecologically speaking? Let's explore this first. When I've told some people about the title of this chapter, they have accused me of being weak. That's right: this chapter is really lame. Some people wanted me to say 'You ARE Living in an Age of Mass Extinction,' as if the 'You may' was the same as 'You are not'.

This in itself is interesting, this understanding of 'may' as 'not'. It has to do with the logical 'Law' of the Excluded Middle. It affects all kinds of areas of life. The normal rule for voting interprets abstaining as saying 'No' when it comes to counting up the votes. You can't interpret it to mean 'Maybe yes, maybe no'. We live in an indicative age, an active one indeed, where a wordprocessing program is prone to punish you with a little wavy green line for using the passive voice; heaven forbid we use the subjunctive, as in 'you might'.

Not being able to be in the middle is a big problem for ecological thinking.

But not being able to be in the subjunctive is also a big problem for ecological thinking. Not being able to be in 'may' mode. It's all so black and white. And it edits out something vital to our experience of ecology, something we can't actually get rid of: the hesitation quality, feelings of unreality or of distorted or altered reality, feelings of the uncanny: feeling *weird*.

The feeling of not-quite-reality is exactly the feeling of being in a catastrophe. If you've ever been in a car crash, or in that minor catastrophe called jet lag, you probably know what I mean.

Indeed, editing out 'may' edit out experience as such. 'You ARE' means that if you don't feel like it, if you don't feel something officially sanctioned about ecology, there's something wrong with you. It should be transparent. It should be obvious. We should deliver this obviousness in an obvious way, like a slap upside the head. 'You may find yourself in' includes experience. In a sense, it's actually much *stronger* than a simple assertion. Because you can't get rid of yourself. You can agree or disagree with all kinds of things – there you are, agreeing or disagreeing. In the words of that great phenomenologist Buckaroo Banzai, *Wherever you go, there you are.*[1]

Philo-sophy

There is something rough and ready about truth, just as there is something rough and ready about philosophy. Philosophy means *the love of wisdom*, not wisdom as such. It's definitely a style of philosophy to delete the *philos* part. There are too many philosophers to mention, and I blush to name them, but you know the type: the kind of person who *knows they are right* and that *you are talking nonsense unless you agree with them*. Needless to say, this is a style I don't like at all. Love means you can't and don't grasp the beloved – that's what you feel, that's what you realize when you love someone or something. 'I can't quite put my finger on it . . . I just love that painting . . .'

Throughout this book, we'll be seeing how the experience of art provides a model for the kind of coexistence ecological ethics and politics wants to achieve between humans and nonhumans. Why is that?

In the late eighteenth century the great philosopher Immanuel Kant distinguished between things and thing-data, as we have begun to see. One reason why you can tell there is a sharp distinction here, argued Kant, is beauty, which he explored as an experience, the kind of moment in which we exclaim 'Wow, that's so beautiful!' (What I'm going to be calling 'the beauty experience'.) That's because beauty gives you a fantastic, 'impossible' access to the inaccessible, to the withdrawn, open qualities of things, their mysterious reality.

Kant described beauty as a feeling of ungraspability: this is why the beauty experience is beyond concept. You don't eat a painting of an apple; you don't find it morally good; instead, it tells you something strange about apples in themselves. Beauty doesn't have to be in accord with prefabricated concepts of 'pretty'. It's strange, this feeling. It's like the feeling of having a thought, without actually having one. In food marketing there is a category that developed in the last two decades or so called *mouthfeel*. It's a rather disgusting term for the texture of food, how it interacts with your teeth and your palate and your tongue. In a way, Kantian beauty is *thinkfeel*. It's the sensation of having an idea, and since we are so committed to a dualism of mind and body – so was

Kant – we can't help thinking this is a bit psychotic: ideas shouldn't make a sound, should they? But we do talk all the time about the *sound* of an *idea*: *That sounds good*. Is it possible that there is some kind of truth in this colloquial phrase?

The German philosopher Martin Heidegger is a controversial figure, because for some of his career he was a member of the Nazi party. This very dark cloud is a big shame, because it prevents many people from engaging with him seriously. And this is despite the fact that Heidegger, like it or not, wrote the manual on how thinking should proceed in the later twentieth and early twenty-first centuries. I hope I'll be able to demonstrate this as I go along, and in addition I hope I can show that Heidegger's Nazism is a big mistake – obviously, but also from the point of view of his very own thought.

Heidegger argues that there are no such things as truth and untruth, rigidly distinguished like black and white. You are always in the truth. You are always in some kind of more or less low resolution, low dpi jpeg version of the truth, some kind of common, public version, *truthiness* (we first met Stephen Colbert's handy term in the Introduction). I know the jpeg analogy doesn't work properly. No analogy works properly. The analogy of truth as more or less pixellated is itself more or less pixellated.

And beauty is truthy. Actually, since I'm not Kant I'm going to say that beauty isn't thinkfeel, it's *truthfeel*. If you want to use the language scientists now use you can say *truth-like*. So if you think about it, we are now at a point where we must acknowledge a subtle flip in our argument.

We've been criticizing factoids as misleading, but why can they be misleading at all? It's because somehow we don't always recognize false things as false. Which means that there isn't a thin or rigid true versus false distinction. In a strange way, *all* true statements are sort of truthy. There is not a sudden point or rigid boundary at which the truthy becomes actually true. Things are always a bit fumbly and stumbly. We are feeling our way around. Ideas sound good. Truthfeel. And you may find yourself living in an age of mass extinction.

The Phenomenon of the Anthropocene

The Anthropocene is the name given to a geological period in which human-made stuff has created a layer in Earth's crust: all kinds of plastics, concretes and nucleotides, for example, have formed a discrete and obvious stratum. The Anthropocene has now officially been dated as starting in 1945. This is an astounding fact. Can you think of another geological period that has such a specific start date? And can you think of anything more uncanny than realizing that you are in a whole new geological period, one marked by humans becoming a geophysical force on a planetary scale?[2]

There have been five mass extinctions in the history of life on Earth. The most recent one, the one that wiped out the dinosaurs, was caused by an asteroid. The one before that, the End Permian Extinction, was caused by global warming, and it wiped out all but a few lifeforms. Extinctions look like points on a time line when you look them up on

Wikipedia – but they are actually spread out over time, so that while they are happening it would be very hard to discern them. They are like invisible nuclear explosions that last for thousands of years. It's our turn to be the asteroid, because the global warming that we cause is now bringing about the Sixth Mass Extinction. Maybe it would make it more obvious if we stopped calling it 'global warming' (and definitely stopped calling it 'climate change', which is really weak) and started calling it 'mass extinction', which is the net effect.

Now it may sound strange, but something about the vagueness of kinda sorta finding yourself in the Anthropocene, which is the reason why the Sixth Mass Extinction event on planet Earth is now ongoing – something about that vagueness is in fact *essential* and *intrinsic* to the fact of being in such an age. This is like saying that jet lag tells you something true about how things are. When you arrive in a very distant strange place, everything seems a little uncanny: strange, yet familiar, yet familiarly strange – yet strangely familiar. The light switch seems a little closer than normal, a little differently placed on the wall. The bed is oddly thin and the pillow isn't quite what you're used to – I'm describing how it feels whenever I arrive in Norway, by the way. Day begins about 10 a.m. during winter. It's pitch dark at 9 a.m. It's still the day, but not quite as you have become habituated to it.

Heidegger's word for how light switches seem to peer out at you like minor characters in an Expressionist painting is *vorhanden*, which means present-at-hand. Normally things kind of disappear as you concentrate on your tasks. The light switch is just part of your daily routine, you flick it on, you want to boil the kettle for some coffee – you are stumbling around, in other words, stumbling around your kitchen in the early morning light of truthiness. Things kind of disappear – they are *merely* there; they don't stick out. It's not that they don't exist at all. It's that they are less weird, less oppressively obvious versions of themselves. This quality of how things seemingly just happen around us, without our paying much attention, is telling us something about how things are: things aren't directly, constantly present. They only appear to be when they malfunction or are different versions of the same thing than we're used to. According to this, you go about your business in the Norwegian hotel room, you go to sleep, and when you wake up, everything is back to normal – and that's how things actually are; they are, as Heidegger says, *zuhanden*, ready-to-hand or handy.[3] You have a grip on them, as in the phrase *Get a grip!* Or the slightly more amusing English version, *Keep your hair on!* (Implying before you quite notice that you are wearing a wig . . .)

Things are present to us when they stick out, when they are malfunctioning. You're running through the supermarket hell bent on finishing your shopping trip, when you slip on a slick part of the floor (someone used too much polish). As you slip embarrassingly towards the ground, you notice the floor for the first time, the colour, the pattern, the material composition – even though it was supporting you the whole while you were on your food shop mission. Being present is secondary to just sort of happening, which means, argues Heidegger, that *being isn't present*, which is why he calls his philosophy deconstruction or destructuring.[4] What he is destructuring is the metaphysics of presence, which is saying that some things are more real than others, and the way they are more real is that they are more constantly present.

Normal for Some, Disaster for Others

This normalization is true – it happens, maybe it does have something to do with sleeping in a place. But is that really because things being handy, *zuhanden*, is the normal state of affairs? Object-oriented ontology is arguing that this ready-to-hand-ness of things is sitting on top of something much deeper and much stranger. There is a weird dislocation between *readiness to hand* and *presence at hand*. Stuff happens without us paying much attention (readiness to hand), yet the same stuff looks peculiar when it malfunctions (presence at hand). This is because things in themselves are ungraspable, totally and completely – irreducibly as they say. Things can't be accessed fully by anything, including themselves. You can flick a light switch, lick it, ignore it, think about it, melt it, fire its protons around a particle accelerator, write a poem about it, meditate upon it until you become Buddha. None of these will exhaust the reality of the switch. The switch could become sentient and develop the power of speech and go on a chat show. What it says on the show wouldn't be the switch – it would be switch autobiography. 'Well, I found myself in the fingers of this philosophy guy, he had jet lag, it was really weird . . . I had a difficult birth.'

Even the light switch would probably say something like

the singer David Byrne in 'Once in a Lifetime' if it ever went on Oprah Winfrey's chat show: 'This is not my beautiful house . . .'[5] And this is because things are mysterious, in a radical and irreducible way. *Mysterious* comes from the Greek *muein*, which means to close the lips. Things are unspeakable. And you discover this aspect of things, as if you could somehow *feel* that un-feelability, in the beauty experience, or as Keats puts it, *the feel of not to feel it*.[6] This 'and you may find yourself' tentative hesitant subjunctive quality isn't just a temporary blip and it certainly isn't just a phenomenon that only occurs to sentient beings, let alone conscious ones, let alone human ones. It's sort of everywhere, because *being isn't presence*.

Kant showed that there's a difference between *the real* and *reality*. It's like the difference between a musical score – a bunch of dots and lines on a page – and the 'realization' of that score by a musician and the audience who showed up to hear it. Reality is, if you like, the *feeling* that it's real: the music is what it is – this is a Bach violin sonata, not a piece of electronic dance music – but it doesn't really 'exist' until you play it or listen to it.

Kant suggests that this 'realizer' is the 'transcendental subject', a rather abstract, universal being that's different from little me, but which seems to follow me around like an invisible balloon, 'positing' things as large or small, fast or slow (it's a pretty boring balloon, only in charge of extension in time and space). Since Kant, a number of other candidates for the 'realizer' have been suggested. Hegel argues that the 'realizer' was what he calls 'Spirit', the grand march of Western human history. Marx argues that it's human economic relations: sure, there are potatoes, but they don't really exist until I've dug one up and turned it into French fries. Nietzsche asserts that it's 'will to power': things are real because you say they are, and you're holding a rifle, so I'm not going to argue.

And Heidegger argues that it's a mysterious being called *Dasein*. The word is German for 'being there', and it's deliberately vague. Heidegger argues that more specific things (such as Kant's 'subject' or the concept of a human or of 'economic relations') are 'modes' of Dasein, a bit like musical key signatures. Ancient Mesopotamia is Dasein in the key of agricultural 'civilization', while the Aborigines are Dasein in the key of Paleolithic hunter-gatherers. Humans don't 'have' Dasein, because Dasein *produces* or *realizes* the human, in the same way that our violinist realizes the Bach sonata. And while there's nothing to suggest that Dasein can't be exclusively human, this is exactly the assertion that Heidegger blunders into. Dasein isn't quite there, constantly – it's a flickering lamplight. But for Heidegger it's exclusively human, and German flickering light is much more authentic than other kinds of flickering light. None of this makes sense. None of it makes sense *on Heidegger's own terms*. This is what OOO is arguing. De-Nazifying Heidegger doesn't mean ignoring him or bypassing him. De-Nazifying Heidegger actually means *being more Heideggerian than Heidegger*.

So if the *truthfeel* of beauty is telling you something true about anything at all – anything at all is called *objects* in OOO, and these sorts of object are sharply different from objectified things, because they are radically mysterious – what truthfeel is telling you is that things are *open*. Also, the beauty experience is telling you that this thing, this thing I can see right here, is ungraspable. It's totally vivid, yet I can't get a grip on it . . . I can't keep my hair on at all. It's like what an American car wing mirror is telling you, out of the corner of your eye: *OBJECTS IN MIRROR ARE CLOSER THAN THEY APPEAR.* Or it's like objects on a shelf by the artist Haim Steinbach. Things are intrinsically kinky, kooky, out of place – this out of place-ness isn't just a function of things breaking and malfunctioning and becoming *vorhanden*. What you experience in jet lag or inside a Haim Steinbach installation is precisely about exactly how things are.

What all this amounts to is that it's the *normalization of things* that is the distortion. A distortion of distortion. Being in a place, being in an era, for instance an era of mass extinction, is intrinsically uncanny. We haven't been paying much attention, and this lack of attention has been going on for about twelve thousand years, since the start of agriculture, which eventually required industrial processes to maintain themselves, hence fossil fuels, hence global warming, hence mass extinction.

Love, Not Efficiency

Restructuring or destructuring this logistics of the world that has grown out of agriculture, which elsewhere I've called *agrilogistics*, is the one thing that would end global warming, but it is usually considered out of bounds, because it implies accepting a non-'modern' view.[7] Agrilogistics means the logistics of the dominant mode of agriculture that started in Mesopotamia and other parts of the world (Africa, Asia, the

Americas) around 10, 000 BCE. Agrilogistics has an underlying logic to do with survival: Neolithic humans needed to survive (mild) global warming, and so they settled in fixed communities that became cities, in order to store grain and plan for the future. They began to draw distinctions between the human and the nonhuman realms – what fits inside the boundary, and what exists outside of it – that continue to this day. They also drew distinctions between themselves (the caste system). Very soon after the agrilogistical programme began, all kinds of phenomena we associate with life in general showed up, in particular patriarchy and social stratification, various kinds of class systems. It's important to remember that these are constructs of history, the consequence of nomads and hunter-gatherers settling down and establishing cities based on a certain form of survival mode.

The modern view was established on (although it thinks itself as a further disenchantment of) now ancient and obviously violent monotheisms, which in turn find their origin in the privatization of enchantment in the Neolithic with its 'civilization'.

Ecological awareness is awareness of unintended consequences. Some ecological politics is about trying to light everything up in a totally nonflickery way, to make sure that there are no unintended consequences. But this is impossible, because things are intrinsically mysterious. So an ecological politics like that would be a monstrous situation, a 'control society', a useful term invented by philosopher Gilles Deleuze to describe our contemporary world. An ecological control society would make the current state of affairs, where kids get tested every five seconds for their ability to resemble a rather slow computation device, look like an anarchist picnic. Even more predictability, even more efficiency. If that's what the ecological society to come will look like, then I really don't want to live in it. And it wouldn't even really be ecological. It would just be this same world, version 9.0.

The ecological society to come, then, must be a bit haphazard, broken, lame, twisted, ironic, silly, sad. Yes, sad, in the sense meant by a character in the British science fiction television series *Doctor Who*: sad is happy for deep people.[8] Beauty is sad like that. Sadness means there's something you can't quite put your finger on. You can't quite grasp it. You have no idea who your boyfriend really is. This is not my beautiful wife. Which means in turn that beauty isn't graspable either, beauty as such – which means that beauty must be fringed with some kind of slight disgust, something that normative aesthetic theories are constantly trying to wipe off. There needs to be this ambiguous space between art and kitsch, beauty and disgust. A shifting world, a world of love, of *philos*. A world of seduction and repulsion rather than authority. Of truthiness rather than rigid true versus rigid false. Truth is just a 1000 dpi kind of truthiness. This isn't the same at all as saying everything is a lie. That's a statement that's trying not to be truthy, which is why it ends up contradicting itself. If everything is a lie, then the sentence *everything is a lie* must also be a lie . . . and so on.

Art That Talks about Its Substances

So we aren't talking about a traditional concept of postmodernity here. In a way, postmodern art, and I'd put Talking Heads' 'Once in a Lifetime' in that category, is in fact the beginning of ecological art, which is to say, art that includes its environment(s) in its very form. Of course, *all* art is ecological, just as all art talks in various ways about race, class and gender, even when it's not doing so explicitly. But ecological art is more explicit. Postmodernism may not have known it consciously at the time, but the ambient openness and strange distortedness of many of its forms talk about the Earth out of which they are ultimately made. Something real is happening. Extreme postmodern thought argues that nothing exists because everything is a construct. This idea, now known as *correlationism*, has been popular in Western philosophy for about two centuries. We just encountered it in our exploration of different kinds of 'realizer'. Again, the idea is that things in themselves don't exist until they have been 'realized', rather like the way a conductor might 'interpret' a piece of music or a producer might 'realize' a screenplay in a movie.

But something funny has happened to this idea. For reality to be correlationist, there has to be a correlatee as well as a correlator: there is a violin sonata, not just a violinist. It's like two faders on a mixing desk. Over time, the correlator fader has been turned way up, while the correlatee fader has been turned all the way down. And this has given rise to the actually rather boring (and definitely anthropocentric) idea that the world is exactly how humans make it, with the correlatee turned all the way down, so down that it sounds like the correlator is doing a solo, not a duet.

The lineage of correlationism starts with Kant, as we saw, who stabilized the explosive idea that causality can't be directly seen, only statistically inferred, the idea with which David Hume blew up pre-modern theories of cause and effect. Kant stabilized the explosion by saying that although causality can't be seen to be running forwards, it can be posited backwards with 20–20 hindsight by the correlator. Again, for Kant the correlator is what he calls the transcendental subject, and since Kant a number of alternatives have been suggested, as I mentioned earlier: the spirit of history (Hegel), human economic relations (Marx), will to power (Nietzsche), libidinal processes (Freud), Dasein (Heidegger), to name a few.

Correlationism is true: you can't grasp things in themselves, facts are different from data, and data is different from things. But that doesn't mean that what gets to decide what's real – the correlator, the decider – is more real than those things, whether the decider is the Kantian subject, Hegelian history, Marxist relations of human production, Nietzschean will to power, or Heidegger's flickering lamplight of Dasein. So while 'traditional' postmodernism, informed by Kant, still relies on this correlationalism, what I'm talking about here, and what underlies OOO, is the idea that this very relationship may not be what we think it is. It may not exist at all.

Dark Ecology

Things are open. Open also in the sense of potential – things can happen in an OOO world, because things aren't totally keyed to human lamplight, they aren't totally meshed together, because in that world nothing could happen, there would just be this completely locked together jigsaw that you could never take apart or put back together. Something happening in one specific place (say a feather falling on pavement) would mean the whole universe changes everywhere. Things are connected but in a kinda sorta subjunctive way. There's room for stuff to happen. Or, as the anarchist composer John Cage put it, 'The world is teeming. Anything could happen.'[9]

So, the strangeness with which we encounter the fact that we are responsible for a mass extinction event is an intrinsic part of it, and not to be deleted. Yelling at people that we are making lifeforms go extinct isn't nice, because it deletes the strangeness. And saying conversely 'Who cares? Everything goes extinct anyway', which is sort of what the right wing often says, and also what some extreme forms of supposedly environmentalist stance say, such as ecological thinker Paul Kingsnorth's Dark Mountain project, isn't nice either, because that also tries to delete the strangeness. This kind of bleak certainty misses how things are.

My approach to ecological thought can be characterized as something I call 'dark ecology'. Dark ecology doesn't mean the absolute absence of light. It's more like Norway in the winter, or the summer for that matter, the way that light in the Arctic reveals something slippery and evanescent about itself, the long summer shadows, the night that lasts for fifteen minutes in Helsinki in June, the dimness. Light as such isn't directly present, you can't pin it down and you can't fully illuminate it: what illuminates the illuminator? Light is splashy and blobby, as quantum theory tells us. And it can't reach everywhere all at once, as relativity theory tells us.

It's like when you die in Tibetan Buddhism. When you die, you see the light – but unlike in some other religions, it's not an obvious light and it's not at the end of a tunnel, and you aren't heading towards it and it isn't the end. In fact, you probably don't notice it at all. It just sort of flickers on, in an incidentally by-the-way sort of a way, and you delete that experience of the nature of mind, then you find yourself being reincarnated. In the traditional literature it lasts for about three seconds, or as the esoteric manuals put it, as long as it takes you to stick your arm into a sleeve three times. You are not deleting some constantly present logos and falling into blurry confusion. In a way you are deleting a wonderful blurry confusion and falling into a fatal certainty.

In Tibetan Buddhism, the time between one life and the next is called the bardo, the 'between'. All kinds of haunting images appear to the consciousness in that state, images based on past actions (karma). We feel that things are different now, that we are in a bardo-like transition space regarding ecological awareness. But really what we are noticing is that things just don't stay put, they don't stay the same. Trying to get over this bardo-like quality results in damage to lifeforms, damage to thinking, damage to experience. The impulse behind racism, for example, is also what empowers a thin and rigid distinction between humans and nonhumans. The violence has already occurred, in the form of the abjection and dehumanizing of some humans. We humans contain nonhuman symbionts as part of the way in which we are human; we couldn't live without them. We are not human all the way through. We and all other lifeforms exist in an ambiguous space in between rigid categories.

If ecological action means *not doing as much damage*, rather than doing things more efficiently, then it's not ecological to insist or slap upside the head or the other similar current modes of supposedly ecological data delivery in general. These kinds of action are like trying to wake us up from this bardo-like dream – but the dreamlike quality is precisely what is most real about ecological reality, so in effect, information dump mode is making ecological experience, ecological politics and ecological philosophy utterly impossible.

Thinking about Groups

Humans have started mass extinction, but me, little me, Tim Morton, and little you, didn't do anything. Once again, nothing, nothing that you did, such as starting your car, has had a statistically meaningful effect. Yet billions of car startings and burstings of coal into flame and so on totally have had an effect. There is an uncanny gap between little me and me as a member of what is called *species*. The human species caused global warming, not the octopus species, let's be very clear about that. But species is exactly what you can't point to. I find that I am and I am not a human, insofar as I did and did not contribute to global warming, depending on what scale you think I'm on, so these scales don't have a smooth transition point between being one human and being part of the total population of humans – suddenly we find ourselves on one scale or another. It's that paradox again. And it seems absurd. Surely seven billion (the current human population) is just one human times seven billion? In computational terms, there is total smoothness between one and seven billion. Yet there is a weird gap.

If you think metaphysically, you can apply a sorites logic to global warming. The sorites paradox is the logical paradox concerning heaps. It's about how vague heaps are – when does a collection of things become a heap? If you take a single rock away from a heap of rocks, does that mean it is no longer a heap? What if you take ten rocks away? Where does the heap start, and where does it end? This quandary suggests a great deal of vagueness, and some philosophers don't like vagueness, so they don't believe heaps exist at all. The trouble is, ecological things such as populations (for example human ones) and ecosystems are very well described as heaps of things. So we had better allow heaps to exist if we're going to be ecological, because addressing global warming and mass extinction can only be done at a massive, *collective* scale.

If you think about it, global warming is a *heap* of actions. Let's analyse it using the logic that results in the sorites paradox. One car ignition firing doesn't cause global warming. Two? No. Three? No. You can work your way all the way to one billion and the same logic will hold. So there is no global warming. Or – drum roll – your logic sucks. How does it suck? It sucks by having no time for things that are in between true and false, black and white. Ecological beings such as lifeforms and global warming require *modal* and *paraconsistent* logics. These logics allow for some degree of ambiguity and flexibility. Sentences can be *kind of* true, *slightly* false, *almost* right.

Heidegger argues that 'true' and 'false' aren't so rigidly different as you might think. You can't delete truthiness without getting into trouble, as I showed a bit earlier, because 'true' applies to the things that Dasein is concerned with, and Dasein is mysterious and slippery. So we are always in the truth, because *Dasein is the truth* we keep trying to seek outside of Dasein. We're always entangled in a thicket of prefabricated concepts that might not apply so well, because of the slippery quality of being. Perhaps this is why social media can be so violent: on Twitter, for example, everyone is trying to be right in one hundred and forty characters or less. Anxieties about 'fake news' exist because in some ways, all news is 'fake'. Everyone is trying to contain or erase the truthiness. But if entities are open, they are not completely nothing, nor are they constantly present, nor are they reducible to other things such as their parts or some access mode such as discourse or economic relations or Dasein. If entities are open, they are truthy through and through. And this actually implies that you can't say just anything you want about entities. You can't say an octopus is a toaster, or that global warming isn't real, or that it wasn't caused by humans, precisely *because* things are open and truthy. Things are exactly what they are, yet never how they appear, yet appearance is inseparable from being, so a thing is a twisted loop like a Möbius strip, in which the twist is everywhere, it has no starting or ending point. Appearance is the intrinsic twist in being.

An agricultural person – aka us – realizing that she is in a twisted historical or ethical or philosophical space experiences what is called tragedy, which is an agricultural-age way of computing the damage caused by an agricultural age. I'm caught in a twisted loop in which my attempt to escape the web of fate has been but a further entwining of that web. Tragedy supposes that looping is evil and that despite the fact that you find you can't escape fate, especially when you try, there is this forlorn hope that in the end, or in some better world over yonder that we can never reach, we might be able to slip those bonds once and for all, hence the ultimately religious horizon of tragedy, where for instance the chorus tells you that there is nothing here that is not Zeus (in ancient Greek playwright Euripides' play *Heracles*).

Tragedy is in fact a small region of comedy space, which is twisted all the way through. Right now, ecological awareness presents itself as tragedy. But sooner or later, we will start to smile, which is maybe how we get to cry for real. Since there is no beyond in which things are indeed totally straight, totally untwisted, it's funny to watch us as a species acting as if there was such a beyond, and constantly slipping into the web of fate, like a slapstick character whose attempt to get from A to B keeps being hampered by his very style of trying to get from A to B. This is why art, which disables getting from A to B by causing the illusion of smooth functioning to malfunction, so as to reveal the spooky openness of things, is in the end joyful and funny, though we need to traverse and respect and not delete a realm of exquisite pain to get there. We really are making this Earth unlivable for ourselves and other lifeforms. I'm not suggesting we just sit back and laugh at that.

Several realms, in fact. Realms of truthfeel. Ecologically speaking, I think the pathway is likely to lead us from guilt down into shame, and from there down into disgust, whence to horror; from there begins ridicule, which dies out in melancholia, whose enabling chemistry is sadness; in turn, sadness is conditioned by longing, which implies joy.[10] At present, the ways in which we talk to ourselves about ecology are stuck in horror mode: disgust, shame, guilt. Eventually things get so horrifying that someone goes 'You gotta be fucking kidding', like that character in John Carpenter's film *The Thing*, looking at the latest mutation of the feminized simulation monster. A ridiculous, absurd laughter breaks out.[11] We aren't quite there yet – we're almost there, which is why some really progressive ecological art, such as the work of the American artist Marina Zurkow, plays with a sardonic kind of eco-humour. We are beginning to trust the tactic of not waking ourselves up from the nightmare, but allowing ourselves to fall further into it, beyond horror. Underneath ridicule space is a melancholy region where things become less horrifying and more uncertain, all kinds of fantasy beings float around like mermaids among the seaweed and submarines. A realm of unspeakable, nonhuman beauty not confined to normative anthropocentric parameters begins to open up.

Another way of saying the same thing is that we are starting to trust that we are in a *catastrophe*, which literally means a space of downward-turning. It's much better to think you are in a catastrophe than to think you are in a disaster. There are no witnesses in disaster. Disasters are what you witness from the outside. Catastrophes involve you, so you can do something about them.

Think about it. This whole 'world without us' fantasy is very suspicious from that point of view. In the last two decades, philosophers and television producers and artists have taken an interest in imagining an Earth without humans. I'm not sure exactly why it started, but I'm pretty sure of the general reason: the media is tuning in to global warming and mass extinction. The paradox is that as you imagine a future in which humans have gone extinct, *there you are, imagining that.* It's a vicarious thrill, like rubbernecking a car accident, and it might be just as obnoxious and dangerous. In the real world, given how entangled we have become with earth systems, if we go extinct it means that many, many lifeforms have also gone extinct or are about to. Opposing anthropocentrism doesn't mean that we hate humans and want ourselves to go extinct. What it means is seeing how we humans are included in the biosphere as one being among others.

This brings up a deep philosophical insight about the fact that we simply can't be on the outside looking in. Scientists call this fact 'confirmation bias' and philosophers call it 'the hermeneutic circle' and 'phenomenological style'. There is no way to escape such things. How I interpret data will depend on what I think I want to find. How I see myself depends on the kind of person I am. How I interpret things is entangled with prefabricated concepts about what interpreting means. This gives rise to a strange insight, which is that living in a scientific age doesn't mean you are living in a cold world of objectivity. It means that you realize you can't achieve escape velocity from your phenomenological style or embeddedness in data interpretation or confirmation bias (three different ways of saying the same thing). We cannot get out.

Funnily enough, living in a scientific age means we have stopped believing in authoritative truth. That kind of truth is pretty medieval, always backed up by the threat of violence because it can't be proved: you just have to believe it. Instead, our modern age is a truthiness domain. Science means we still might be wrong, and we may find ourselves holding on to a bunch of weird assumptions that don't quite make sense, but this is better than firmly believing we are right because the Pope ordered us to believe whatever.

Mass extinction is so awful, so incomprehensible, so horrible – and at present it's so invisible. We hardly know where to start, apart from either ignoring it or electroshocking ourselves about it. One of the recent mass extinctions, the End Permian Extinction, also involved global warming. It happened about 252 million years ago, and at that time, plants were to blame. Unlike plants, we can choose not to emit excessive amounts of carbon, so it's not inevitable this time.

When I say *recent*, I'm alluding again to the fact there have only been five previous mass extinctions in the four-billion-year history of life on this planet. That fact alone, that fact of deep time, is horrifically disturbing. It was disturbing in the early nineteenth century, when geologists began to figure it out, and it's disturbing now. We used to tell ourselves that it was disturbing to the poor dumb Victorians because it shook their faith in God. In exactly what is it shaking our faith now?

Ecology without Nature

Ecological awareness is shaking our faith in the anthropocentric idea that there is one scale to rule them all – the human one. Nietzsche announced that God was dead in the nineteenth century, and this is often taken to imply that humans face a meaningless existence. But this isn't true. It's the opposite. The death of God isn't some empty, desolate wilderness, it's a scary jungle swarming with creatures – literally. It's thousands of equally legitimate spatiotemporal scales that have suddenly become available and significant to humans. We are so habituated to living and thinking on a very small range of timescales that students who train as geologists say that they have to go through a process of acclimatizing to much vaster tracts of time.

Now we know that ecological awareness means thinking and acting ethically and politically on a lot of scales, not just one. It's not true, however, that this will feel like the kind of powerful thrill you get from playing with one of those online scale tools that zoom you in and out from the Planck length (the smallest currently measurable one) to the scale of the entire universe, or those humbling-yet-empowering clock faces on which humans appear at the last second before midnight; or those floor diagrams some scientist presenter walks across to show how we appear at the last sliver on the bottom right-hand corner. The scale in all of those is smooth and consistent – it's a sort of hollowed-out, blown-up version of the good old anthropocentric scaling, only now we are in a privileged godlike position of omnipresence outside the universe, where every scale is just a toggle away. But it isn't like that at all. That kind of thing confuses time with the *measurement* of time, and further it confuses the measurement of time with just a *few* kinds of measurement – the kinds that are convenient for humans. It's not just true that there is a time for everything, as it says in Ecclesiastes ('a time to reap and a time to sow . . .'); it's the case that from grasses to gorillas to gargantuan black holes, *everything has its own time*, its own temporality.

Psychological research has shown that we are good at narrating the correct sequence of geological events: Earth emerges from a cloud of dust and gas, microbes evolve, followed by sponges, fish, butterflies, primates . . . But very few of us are able to imagine the right *durations* of geological time without special training. And being able to understand durations is particularly important for us right now, because global warming's effects may last up to 100,000 years. What does that actually mean? We tend to have only two vague temporal categories in our heads: ancient and recent. We use these as a template to conceptualize what we call 'prehistory' (the pre-'civilization' human stuff, and the nonhuman stuff) and 'history' (the 'civilization' stuff). It would be better, more logical and requiring fewer beliefs to see everything – even now – as history and to see history as not exclusively human.

I think we have more in common with the Victorians than we'd sometimes like to admit. Indeed, the decisive emergence of what I call *hyperobjects* on our radar makes the sensibility of our contemporary moment extremely Victorian. Mary Anning discovered a dinosaur skeleton in an English cliff face, and the abyss of deep time opened up. The vast distributed processes of evolution were discovered. The gigantic Pacific weather system El Niño was discovered later in the nineteenth century. Marx traced the invisible workings of capitalism. Freud discovered the unconscious. And once again we stand in awe of gigantic entities massively distributed in time and space, in such a way that we can only point to tiny slices of them at a time. Once again we find our faith shaken, and now it has clearer contours: it's not about the disappearance of an agricultural-age god. It's much, much worse. It's about the flip side, the unconscious, the unintended consequences of our faith in progress, which far precedes agricultural-age gods, as a matter of fact, and is their condition of possibility. A 12, 500-year-long social, philosophical and psychic logistics is now showing its colours, and they are disastrous.

And for the longest time these logistics were called Nature. Nature is just agricultural logistics in slow motion, the nice-seeming buildup to the Anthropocene, the gentle slope of the upwardly moving rollercoaster that you don't even suspect to be a rollercoaster. Agricultural society coincided with the Holocene (our current geological period, which started over 10,000 years ago, marked by the retreat of the glaciers), which was remarkably stable and cyclic as far as Earth systems such as the nitrogen and carbon cycles went. It's controversial, but some geologists actually think that the periodic, smoothly cycling form of the Holocene was in fact a product of the functioning of a certain agricultural mode. This mode began in Mesopotamia and elsewhere on Earth at the start of the Holocene. If it's true that agriculture contributed to the stability of Earth systems, it makes things even more disturbing. Like when someone has a seizure, and their brain waves become beautifully regular just beforehand. Or before an earthquake, when the same thing happens to the tectonic plates. On this view, what is called Nature – the smooth cycling represented so nicely in feudal symbolic systems – is directly the Anthropocene in its less obvious mode. Then comes the huge Earth systems data spike we see in the ex-American Vice President Al Gore's movie *An Inconvenient Truth*, the spike that starts around 1945, evidence of runaway carbon emissions.[12] Everything starts to go haywire.

The inner logic of the smoothly functioning system – right up until the moment at which it wasn't smoothly functioning, aka now – consists of logical axioms that have to do with survival no matter what. Existence no matter what. Existing overriding any *quality* of existing – human existing that is, and to hell with the lifeforms that aren't our cattle (a term from which we get *chattels*, as in women in many forms of patriarchy, and the root of the word *capital*). Existence above and beyond qualities. This supremacy of existing is a default ontology and a default utilitarianism, and before any of it was philosophically formalized, it was built into social space, which now means pretty much the entire surface of Earth.

You can see it in the gigantic fields where automated farm equipment spins in its lonely efficient way. You can feel it in the field analogs such as huge meaningless lawns, massive parking lots, supersized meals. You can sense it in the general feeling of numbness or shock that greets the fact of mass extinction. Quite a while ago humans severed their social, philosophical and psychic ties with nonhumans. We confront a blank-seeming wall in every dimension of our experience – social space, psychic space, philosophy space.

Uncannily we begin to realize that we are somewhere. Not nowhere. And we may find ourselves living in an age of mass extinction. I'm all for letting us linger in the strange openness of this uncanny discovery that space was just a convenient white Western anthropocentric construct for navigating your way around Africa to reach the Spice Islands, and so on. Because strangely, this feeling of openness, this uncanny sensation of finding ourselves somewhere and not recognizing it, is exactly a glimpse of living less definitively, in a world comprised almost entirely not of ourselves.

What then can we say about this world? How do we talk about it? What does the fact of ecological interconnection mean? We're going to find out in the next chapter.

CHAPTER 2

...And the Leg Bone's Connected to the Toxic Waste Dump Bone

'Everything is connected.' You hear phrases like that a lot when you talk or read about ecology. But what does it mean? It sounds easy to understand, but actually, it's quite strange. When we consider ecology, we find that things are even more connected than we might assume. And that even more weirdness results when we start to let in this deeper connectedness.

For example, we often hear about something like 'the fragile web of life'. And when we hear this, maybe on a TV documentary, we nod sagely and go 'Yeah, I know. The fragile web of life. That thing.' It's like being in church listening to a sermon you don't quite understand, but you feel this group pressure to nod along. Or you hear things like 'He's got the whole world in his hands' (the hymn) or 'I'd like to buy the world a Coke' (the advertisement). Or you see one of NASA's 'blue marble' *Earthrise* photos.

All these experiences are aesthetic. They are about how things look or feel. They are neither true nor false. In other words, when we visualize these sorts of things, we don't know what we're talking about. We *think* we do. This must mean that there are a host of untested, unexamined ideas and beliefs structuring these sorts of well-known images of

Earth. For example, one thing we tell ourselves is that the blue marble sorts of photos show us a world whose precious wholeness contains us like tiny fragments. But what is this wholeness really, and are we actually parts of it, and what kind of part? If you thought it all sounded vaguely religious, you'd be right. A lot of thinking ecologically sounds religious, because it involves extremely profound and hard to express (at least at present) concepts and feelings. But it's also related to religion because religion as we know it arose during the agricultural period we call the Neolithic Era, and this period structures our world, and the structuring is responsible (now that it's persisted for 12,500 years) for our ecological crisis. So we had *definitely* better examine religion.

Is there an end to the song we derive from Ezekiel, about the bones? I mean, Ezekiel, agricultural-age religious exponent as he is, wants the parts to be reconnected into a wonderful whole, by God. But can you actually stop the explosion of 'is connected to'-s? Think of a dictionary. The meaning of a word is a bunch of other words. And so on: you look those words up in turn. You keep going. What do you think will happen? Will you arrive back at the first word in a nice neat circle? Or will your journey look more like a tangled spiral? Even if you made it back to the first word, by chance, would that look circular? I don't think so. And I think the same thing happens when we consider how lifeforms are interrelated.

Things and Thoughts

There is a really deep reason why, when you examine things from an unusual (to humans) point of view, they become strange in such a way that you need to include your own perspective in your description, as if you were like Neo in *The Matrix*, touching the mirror only to find that it is sticking to your finger and pulling away from the wall as you try to withdraw your hand.[1]

It's like what happens in a dream. When you dream of nasty creepy-crawlies falling on you from the ceiling, you also have a certain feeling or attitude (or whatever you want to call it) towards the insects, perhaps horror or disgust, perhaps mixed with a strange detachment. This is the same as how in a story there is what's happening (the narrative) and how it's being told (the narrator, whether it be singular, plural, human or not, and so on). These two aspects form a *manifold*. When we look at a 'thing', we are forgetting that 'thing' is just part of a manifold. It's not true that there's 'me' and then there's a 'thing' I reach out to with my perception, like reaching my hand out to a can of beans in the supermarket. But perhaps we have tried to design our world to look like a supermarket, full of things we can reach out and grab.

The result of living as though you believe in subject – object dualism, which is our usual mode of thinking about the world (even if we are doing it unconsciously), is that it becomes hard to accept what is in fact more logical and easier on the mind in the end. When you analyse a nightmare, you discover that the insects and the feelings you are having about them are both aspects of your very own mind. Perhaps the insects are unacceptable thoughts of which you're just becoming aware. What is so powerful about psychoanalysis and some spiritual traditions such as Buddhism is that they enable you to entertain the idea that thoughts and so on are not 'yours' all the way down, which can be very liberating: what matters isn't exactly *what* you think, it's *how* you think. You know that facts are never just 'over there' like cans of soup waiting to be picked up in some neutral way. You know ideas code for attitudes, insofar as ideas always imply a *way of thinking* them, an attitude, and that this explains how propaganda works. Take a very simple example: the term *welfare* evokes contempt for its recipients in a way that the word *benefits* doesn't. Since 2010 the British Conservative Party succeeded in getting almost everyone in the media to say 'welfare' and not 'benefits', with the obvious repercussions of making cuts more acceptable. Reading a poem is a wonderful exercise in learning how not to be conned by propaganda, for this very reason. That's because a poem makes it very uncertain exactly what sort of way you are supposed to hold the idea it presents. If I say 'Come here!', it's fairly obvious what I mean, but if I say 'It is an Ancient Mariner', you might be a bit flummoxed. Reading a poem introduces some wiggle room between ideas and ways of having them. Propaganda closes this space down.

Something fascinating occurs if you start to think how the biosphere, as a total system of interactions between lifeforms and their habitats (which are mostly just other lifeforms), is also like the inside of a dreaming head. Everything in that biosphere is a symptom of the biosphere. There is no 'away' that isn't merely relative to a certain position within it. I can't suppress my thoughts without them popping up like nasty insects in my nightmare. I can't get rid of nuclear waste just by hiding it in some mountain. If I widen my spatiotemporal scale enough to include the moment at which the mountain has collapsed, I didn't really hide the waste anywhere once and for all. You can't sweep things under the carpet in the world of ecological awareness.

And this biosphere includes all the thoughts (and nightmares) we are having too. It includes wishes and hopes and ideas about biospheres. It's not exactly physically located precisely on Earth. It's *phenomenologically* located in our projects, tasks, things we're up to. Say, for example, we decide to move to Mars to avoid global warming. We will have to create a biosphere suitable for us from scratch – in a way we will have exactly the same problem as we have on Earth, possibly much worse, because now we have to start from the beginning. Experientially, which is a sloppy and biased way of saying the philosophical word *phenomenologically*, we are still on Earth. Sloppy and biased, because it implies all kinds of things that need to be proved in turn, such as the idea that there is a certain kind of 'objective world' and that 'subjectivity' is different from it. The phenomenology of something is the logic of how it appears, how it *arises* or *happens*. If we move to Mars, the move will appear in an Earthlike way, no matter what the coordinates on our space chart tell us.

So it's not correct to say that the biosphere is 'in' a preexisting space. The biosphere is a network of relations between beings such as waves, coral, ideas about coral and oil-spewing tankers, a network that is an entity in its very own right.

As the systems theorist Gregory Bateson implied when he wrote about 'the ecology of mind', mental issues are somehow ecological in this sense.[2] How your thoughts are related equals what is called 'mind', and mind is like the biosphere. Even though it's made up of thoughts, mind is independent of those thoughts, it affects them causally. If you are scared, you will think scary things. It's what some people call 'downward causality'. Something like climate can affect something like weather. It's not true that climate is just a graph of how weather events are related. There is something real there. You can't *reduce* the biosphere to its component parts, just as you can't reduce your mind to its component thoughts. And you can't reduce your thoughts to what the thought is about, or to the way you are thinking about that thought: you need both, because a thought is a manifold. And this leads to a very interesting insight: maybe *everything is a manifold*. Or to use Bateson's language, a 'system'. The system is different from the things out of which it is made. Being mentally healthy might mean knowing that *what you are thinking* and *how you are thinking* are intertwined.

It's not exactly *what* you believe but *how* you believe that could be causing trouble. In other words, there are beliefs about belief. Maybe if we change *how* we think about things such as coral and white rhinos, we might be more ecologically healthy. And maybe mental health and ecological 'health' are interlinked. I believe that humans are traumatized by having severed their connections with nonhuman beings, connections that exist deep inside their bodies (in our DNA, for instance; fingers aren't exclusively human, nor are lungs or cell metabolism). We sever these connections in social and philosophical space but they still exist, like thoughts we think of as unacceptable and that pop up in nightmares.

Part of our growing ecological awareness is a feeling of disgust that we are literally covered in and penetrated by nonhuman beings, not just by accident but in an irreducible way, a way that is crucial to our very existence. If you didn't have a bacterial microbiome in your digestive system, you couldn't eat. Maybe this feeling of disgust will diminish if we become used to our immersion in the biosphere, just like our neurotic feelings diminish as we become friendlier with our thoughts – perhaps through psychotherapy or meditation. There have indeed arisen forms of ecological psychotherapy, and a branch of psychological studies some call ecopsychology. And many Buddhist meditation teachers also write about ecology, as a glance at some of the readily available magazines such as *Shambhala Sun* will show you.

Mashed Up, or Exactly How Much Connection?

So by now I hope you have started to look at the big picture, or what some philosophy calls *totality*. Or what some meditation manuals call *panoramic awareness*. And panoramic meditative awareness is unique and specific. It's not just a colourless flavourless odourless box with thoughts churning around in it. It's more like an electromagnetic field with a specific frequency.

What does this mean? Being a bit more aware or enlightened doesn't mean becoming omniscient or omnipresent, or for that matter the inverse, becoming a stupid zombie that can't even brush her teeth or answer the phone. Buddha can drive a car and knows how to flush the toilet. In the same way, like the hum of a huge orchestra, a biosphere has very specific qualities that can't be reduced to the parts of the biosphere.

How everything is interconnected is also a *thing*.

The fact that interconnection is also a thing, not just an abstraction or convenient idea, has really surprising, deep implications. But in order to examine them we will need to take what seems to be a bit of a detour. Bear with me while we start out. We need to begin by considering things that we often call ideas.

It's not what you think but *how you think* that starts World War III. This is true in Buddhism, and in William Blake's poetry: his *Songs of Innocence and of Experience* are all about what he calls 'contrary states of the human soul', which we could also describe as 'different modes of thinking about believing'. To a hammer, everything looks like a nail. To a cynic, everything looks hopeless and hopeful people look like fools. So you can lie in the form of the truth. You can say 'We're totally screwed' in a way that contributes to being totally screwed, because you disempower your listener through cynical reason. Plenty of environmentalist speech gets stuck this way. That's another way in which 'Earth is dying' is not a helpful thing to say at all, even if it's somewhat true. It's quite sensible – and ecological – to resist such jeremiads, and this doesn't mean you support big oil corporations. You have started thinking about how you think as part of a dynamic manifold that includes what you are thinking about, along with things that aren't just thoughts, such as forests and cities.

It only sounds hippy-dippy and weird because we're not used to it, because we've been bankrolling the agricultural projects that eventually resulted in the industry that leads to global warming, bankrolling with all our philosophical, psychological and spiritual might. According to the inner logic of how we go about our business, things are objectified lumps of something like plastic, lying 'over there', that I can manipulate at will. A huge amount of violence goes into sustaining this view, precisely because it isn't accurate. Once again what we think and how we think it are deeply connected.

In Western philosophy, it was the German phenomenologist Edmund Husserl who started us off thinking in this 'manifold' way. The years right around 1900 were very significant for developments in science (just think of relativity theory). Yet it was also the moment when an earthquake happened in Western philosophy. Husserl reasoned that ideas don't just float around in space, but are instead what are called *phenomena*: they always have some kind of colour or flavour, and this colour or flavour isn't a decoration or an optional extra, but intrinsic to what an idea is. Among other things, Husserl was reacting to a movement in logic in the nineteenth century called *psychologism*. Psychologism argued that logical sentences were symptoms of a healthy brain. In other words, making logical sense was derived from a brain that was functioning properly (whatever 'proper' means). Logical sentences are sentences such as *If p, and if p then q, then q*: given the fact that there are bananas, and since if there are bananas then there are banana trees, in that case there are banana trees. They make sense, says psychologism, because healthy brains make them. But what is a healthy brain? Well, it's a thing that can make a logical sentence. And what is a logical sentence? Well, it's a thing that comes out of a healthy brain. And what is a healthy brain? We will need some kind of science to verify what a healthy brain is to break this vicious circle. But science relies on logical sentences. And what is a logical sentence? It's a symptom of . . . and so on. There is an infinite regress at work and we haven't actually said anything at all.

So, reasoned Husserl, this just can't be how things are. Logical sentences can't be just symptoms of something.[3] We can't *reduce* them to being the output of healthy brains. They have a reality all their own. Instead of being evidence of proper mental functioning or, to extend this thought, even of proper human DNA, whatever that is, logical sentences have their own building blocks, their own DNA. And they can manage on their own. A logical sentence is like a Tweet or a meme: it has its own sort of life, which means that it's distinct and unique – it has a colour and a flavour and a texture. Like a hammer, you have to handle it this way, not that way.

Husserl's understanding was like finding that an ocean, far from being vast and empty and bleak, was swarming with fish. What was the ocean? The ocean of reason that Kant had established slightly more than a hundred years earlier. Kant upheld that it doesn't matter to what he called pure reason that little me, Tim Morton with his specific size, shape, colour and gender, wishes and hopes and so on, is thinking reasonable things. There is something *transcendental* about reason. You can't point to it, but it's real. This ocean of reason sort of floats just a little bit behind my head. It's a rather cold, uninhabited, eerily clear ocean, because it just does one thing: it mathematizes, measuring things and telling me that this galaxy is *this* big and has lasted *that* long and has *this* kind of movement through the universe. But Husserl showed that because logical sentences have a reality all their own, other types of sentences do too, such as hopeful sentences, wishing sentences, hating sentences . . . It was as if Husserl had discovered that the Kantian ocean had all kinds of differently coloured fish swimming in it, fish with their own DNA structure independent of little Tim and Tim characteristics such as having reddish facial hair. Kant had shown that there was a very significant part of reality that you couldn't point to – the ocean of reason – and Husserl then showed that this ocean is inhabited after all, and that the fish that swim in this ocean are entities in their own right, with their own DNA.

And these fish aren't just restricted to propositions that look logical to the untrained eye. There are all sorts of logical fish, as well as hoping fish, loving fish, hating fish, imagining fish. These are all *intentional objects*, *intentional* meaning that they are contained within this thought-ocean ('intentional' here means 'held within the mind', not the usual sense of 'pointing at some external goal via some mental act'). Just as there's a certain way to handle a shark, there's a certain way to handle a feeling of disgust – there is a mode of having that feeling that goes along with the feeling. And like a magnet, the shark and shark-handling mode are two poles of a phenomenon: they go together, in an inextricable way. Which means that it's not quite true to say that 'I' am 'having' a 'thought'. It's more like this: 'I' is something I sort of deduce or abstract from the phenomenon of this particular thought, just as what the thought is about is also part of that phenomenon.

We are so used to thinking in a dualistic way, that the implications of the fact that thoughts are independent of the mind sound unbelievable. But it's pretty hard to push Husserl's insight over, because just as in Kant, it doesn't depend upon believing something external to the argument; there is no other ecology outside of the one you're currently in, examining this argument. Phenomena don't just happen, then you perceive them. The phenomenon *includes* the act of having it, hammering it, measuring them, mathematizing it, feeling it.

And in turn this means something rather amazing about activities like hammering. A hammer is a certain something, a very specific something – and yet it's not a hammer exactly. It's all kinds of things to all kinds of beings. It's a landing strip for a fly. It's a surface for dust to collect on. It's a hammer when I start using it for my hammering project. But a hammer doesn't just wait around in outer space for someone to grab it. Hammers happen when you grab a metal-and-wooden thing for hammering in a picture hook. In this way a hammer is like a poem. A poem isn't the squiggles on the page. It's how I orchestrate those squiggles when I read them, how an editor interprets the poem by putting it next to some other poems in an anthology, how the poem is taught in a poetry class.

The World is Full of Holes

Hammering is a very vivid, specific thing with its own DNA, which includes me and my wish to hammer in this picture hook, a metal-and-wood thing called 'hammer', the wall, the hook . . . the hammer bone's connected to the wall bone . . . So the full-on, twelve-inch remix of Husserl is full-on object-oriented ontology, in which things are not exhausted by how you use them; they don't hang around in outer space waiting for someone to use them, interpret them, hammer with them. Things are not *underneath* how they appear, where 'appear' means something really general that includes being part of phenomena such as *eating*, *hammering*, *interpreting*, *reading* . . .

There is always some kind of truthy interpretation space in which your thoughts and ideas and actions are taking place, and the thing to remember about this space is that (1) it's not optional and (2) it's not totally sealed off, it's perforated. What does that mean? First of all, it means that not only the mental but also the physical (and psychic and social) ways we 'interpret' things are in that space. A violinist interprets Berg's violin concerto when she plays it. When I hammer in this picture hook, I am interpreting the wall in the key of hammer. And the hammer relates to the wall, which relates to my house, which relates to the street, which relates to the drains in the street, and so on . . . Is there any way to stop the explosiveness of this context, physical and non-physical, in which what I'm up to is taking place? Why, no.

Thoughts and statements that try to achieve escape velocity from their embeddedness in interpretation space just can't do it at all. When you jump outside the world in order to judge it (whatever world it is: poem, Michael Jackson video, plant, Earth), *there you are, doing that*. This is not a superficial fact. It means that the search for a perfect *metalanguage* that would act as the perfect policeman for all the other 'object' languages is impossible. It's like in that famous Monty Python sketch, 'The Argument Clinic'. A man walks into an office and says that he wants an argument. The bureaucrat behind the desk refuses. Then they argue about whether this is the beginnings of an argument or not. Then a policeman arrests them both under the 'Silly Sketches Act'. Then another policeman arrests all of them, including the first policeman. Then another one comes in and arrests everyone else. The scene ends with another policeman's hand clapping the shoulder of the last policeman[4] . . . *All of them, all the policemen, are part of 'The Argument Clinic' sketch.*

That's number (1). And number (2) is related, in a strange way. Whatever world I'm in is never complete and it's never totally mine (I myself am never totally mine). I can't add a royal seal or a special sentence or a floral wreath or a cherry on top of the cake that would guarantee that it was self-identical all the way through. This is great news, because it means that the notion of *world* is *perforated*: I can share my world with a tiger and the tiger can share her world with me. Our worlds can overlap. Heidegger argued that only humans have a full-on, rich 'world', while lifeforms that wriggle around ('animals') are 'poor in world' and things such as stones have no world at all. There is no reason for him to assert this and, double trouble, his assertion means that for him, *world* is totally sealed and solid – which on the basis of his own theory, where things can't be grasped directly, just can't be true. Nazism for Heidegger was a way for him to cover over and ignore and keep anthropocentrically safe from the most radical implications of his own theory. In fact, many Western philosophers since – and including – Kant have sought refuge from the outer-limits weirdness of their own theories in all sorts of ways, to avoid them outstripping what they think they want their thoughts to do – maybe they are afraid of what people might think, or maybe they're just afraid of what they might think, which could be out of step with how they otherwise live their lives.

I could swap out this thing that the hardware store calls a hammer for this thing that we call *lump of wood* or this hardened elk sausage over here . . . and use that on my hammering mission. I'm still hammering, though it might not work as well if the sausage meat isn't hard enough, and maybe it'll leave a bit of a greasy stain on my wall. Never mind: at least I've shown that there is a sharp difference between things that call themselves, or are called, or look and quack like hammers and the 'hammering mission' that includes walls and pictures and me wanting to hammer things and me wanting to impress my dinner party guests with my nice new picture. Phenomena such as hammering don't work unless there are things that might also be landing strips for flies, with lumps of metal at one end, things that aren't exhausted by my hammering mission. It's as if below the top level of Husserl's ocean, in which the fish are swimming about – fish such as hoping and wishing and planning to hammer in a picture – there is a sparkling coral reef of all kinds of things that the fish depend on. It's just that they're not your granddaddy's thing. They aren't quite capable of being pointed at directly, because pointing at is also a mode of access, just as good or bad at accessing things as hammering is. Pointing to doesn't exhaust what things are either.

And this coral reef of things includes the biosphere.

The biosphere isn't just a convenient label for a whole bunch of things that join together. The biosphere isn't just a context that appears because I'm interpreting that bunch of things a certain way. The biosphere is its own unique, distinct thing, and this thing is distinct from its parts, which include trees and worms and coral and ideas about biospheres. So the big picture here is that your ideas about, and your feelings about, and your plans about, lifeforms and the biosphere coexist along with lifeforms and the biosphere. They are part of what is connected together. You aren't outside the biosphere looking in. You are glued to it, in a way that's much more super than Super Glue.

How so? You are glued to the biosphere *phenomenologically*. This means that even if you are physically far away from it – if you could take a tape measure and figure out that you were 200,000 kilometres away, outside Earth's gravitational field, you would still be 'in' it in a phenomenological sense, based on the kind of philosophical argument we have just been exploring. Just to repeat something I was saying earlier, imagine you want to move to Mars, because the biosphere down here on Earth is in really bad shape. On Mars, you will have to recreate the biosphere from scratch – in a way you have an even worse problem than the one you faced on Earth. Even though empirical measurement is telling you that you are millions of miles away, you are still there, on Earth.

It seems right, but it's just not accurate, to say that things are like products on a supermarket shelf, and that you reach out towards them from some nebulous place inside you called 'mind' or 'self'. Think of Saturn. You are on Saturn, right now – part of you is, anyway. You are thinking, in Saturn mode. Thinking, in the key of Saturn. Your mind is wherever you put it. You are 'in' the biosphere in a much more powerful sense than Google Maps points out when it locates you 'in' a particular street. You are 'in' the biosphere in the sense of being 'into' it: you are concerned about it; you care about it. You are locked together with the thing you are concerned about. You form a unit, no matter how spatially close or far you are from one another: you are phenomenologically *near* even if you are on the other side of the Galaxy.

The Mesh: Where Do You Draw the Line?

What we want to do and how we feel and what we are wanting and feeling about are all mashed together. Now let's examine this mash, which elsewhere I've called *the mesh*.[5] Here is a rather official-sounding philosophical way of putting it: the context of relevance is *structurally incomplete*. The mashup isn't ever a nice neat complete circle. Whenever you want to do something, you always encounter a whole thicket of things that are relevant to what you're wanting to do. You want to get to the supermarket, so you need your car, which requires the road, which means there need to be highway regulations, which rely on city councils, which have to do with fixing potholes, and these potholes snag your wheels as you try to get to the supermarket . . . and so on . . . You will find that you won't be able to stop the explosion of contextualization.

Ecological awareness is another name for this context explosion. Leg bones aren't just measurably connected to hipbones, and these aren't just measurably connected to toxic waste dumps. They all have to do with each other, and this loop of having-to-do isn't a nice neat circle at all, but a sprawling lasso that seems able to gather *everything else* into its loop. Normally we try to contain or curtail the lassoing business. But ecological awareness means you have started to allow the lassoing to go on and on, possibly for ever.

The amazing conclusion is that no neat circle of context will ever fully explain the thing you're trying to explain. Ecological awareness gives you a world in which everything is relevant to everything else, but is also really unique and vivid and distinct at the very same time. In this world, everything you think and feel is *relevant*, in the 'leg bone's connected to the toxic waste dump bone' way I've been using that term.

And in turn, this means that feelings of indifference are also relevant. Feeling disconnected from ecological awareness is another mode of . . . ecological awareness.

This should be fantastic news, because it means that ecological awareness is now really cheap. You don't have to cook yourself into a special state of mind to have it. You don't have to completely transform the world in order to be ecological.

You don't even need to make your relevance lasso wider. Just the mere idea that there is a relevance lasso is enough to make you notice. Because *world* is always a bit ragged and broken, because the lasso is never nice and neat and circular, your mission can connect and interact with the missions of others. And because this ragged, broken world is quite cheap, all kinds of beings can have a world, whether we think they are intelligent or even conscious or even sentient. A butterfly can have this sort of ragged, not-completely-closed world, with all sorts of jigsaw pieces missing. A tree can have it.

Being-connected-to is not as big a deal as the very high-minded eco people make it out to be. When they make it out to be a big deal, they are setting the bar for ecological awareness really high. As though being ecologically aware is like being enlightened, or purifying one's sins, or like being capable of seeing everything and everywhere all at once. But I hope we've put to rest the oppressive possibility that you *can* see everywhere at once. And since you can't see everywhere at once, you can't ever grasp the whole, because wholes aren't actually like that – they aren't everywhere, they don't fit over everything. The members of wholes are always in excess of those wholes.

This means that ecological awareness and ecological action are much easier than we have been thinking. *You are already having ecological awareness and doing ecological action, even by ignoring or being indifferent to them.* Once you figure this out, things become much easier, at least easier on your mind and heart. You have some wiggle room, because relevance has wiggle room, because things have wiggle room, because things never quite coincide with how they appear for or how they are used by or interpreted by other things (and possibly even themselves).

OK. If things are related in a way that isn't about coming under an umbrella that is always greater than them, where do you draw the line? Because you have just decided that there is no umbrella big enough to contain everything, because there's always more of everything than there is umbrella.

You have just got up to speed with the title of this chapter. You have arrived at a way of organizing things based on what some philosophers call *contingency* and what some linguistics call *metonymy*. You have, in other words, started literally to say to yourself, 'The leg bone's connected to the hip bone. And the hip bone's connected to the chair bone. And the chair bone's connected to the prison chair factory bone. And the prison chair factory bone's connected to the toxic waste dump bone. And the toxic waste dump bone's connected to the biosphere bone. And the biosphere bone's connected to the bower bird bone (you see, you can go smaller as well as bigger). And the bower bird bone's connected to the rainforest bone. And the rainforest bone's connected to the electromagnetic shield around Earth bone. And the electromagnetic shield around Earth bone is connected to the spinning iron core of Earth bone. And the spinning iron core of Earth bone is connected to the supernova-where-the-iron-was-formed-bone . . .' This interconnection without an edge or centre is what French philosopher Georges Bataille calls a *general economy*.[6] And 'economy' here doesn't just mean 'how people deal with money'; it means how people organize their enjoyment, how they exchange and circulate things (and so on). Thinking of economies – systems of interrelating actions – as 'restricted' or limited, the way we often think of the cycles of life or the water cycle or in the ideas we have about recycling – are always open to this more general, unravelled possibility space. That's because closed systems must inhabit some larger, less organized space in which they can assume a variety of different states, but not every single possible state. Restricted economies are like lumps in your custard. They are made out of custard and can easily collapse back into the softer general custard mixture.

Where on earth (and in heaven, for that matter) can this explosion of connections stop, and does it matter if it can't?

What we are talking about is an explosion of context: a *context explosion.* What's interesting is that most contextual criticism in the humanities seems hell bent on *containing* this explosion. Scholars will explain (often explain away) a cultural artefact by relating it to the decade and the country or the circle in which it was composed; or the biography of the author; or the state of (human) economic relations at the time in the author's country. All of this is important information for understanding things, people and events, but it does not represent the limits of knowledge or understanding. We cannot ever reach a total understanding of even a single book, thought or painting, no matter how much information we amass. By comparison, ecologically aware criticism opens up a vertigo-inducing abyss of potentially infinite, overlapping contexts. So that by definition, there can be no one context to rule them all.

Not Your Grandaddy's Holism

We're getting a handle on the 'web of life' type of image, only in the previous section we've beefed it up with some kind of logical backbone. And now that we've done this, we are in a position to understand that we can't really, not with a straight face anyway, reduce the whole – the 'thing' formed by the interconnections – to its parts. But we also find out that we can't reduce the parts to the whole. 'Reduce' doesn't mean 'break into smaller bits'. Physical wholes are obviously bigger than their parts. What we mean by 'reduce' is 'explain away in terms of something we consider to be more real'. What this means is that – wait for it – *the whole is always less than the sum of its parts.*

Wait a second. This is crazy! Haven't we been telling ourselves, all our lives, that the whole is always *greater* than the sum of its parts? Isn't that the point of the blue marble photos? That if we don't care for Earth as a whole, all the little things squiggling around on its crust will vanish? And doesn't this mean that Earth is more important, in fact *more real*, than the squiggling things (blue whales, humans, slime moulds)? What's happening?

What's happening is that we're using logic in order to stop retweeting something we keep saying that we've never proved: that the whole is always greater than the sum of its parts. It has always sounded really mysterious to me, but somehow we keep saying it as if it's true. This is a belief, and it affects so many things. We might think of consciousness as something that emerges from the 'hum of its parts' as it were, the operation of all the brain firings. It's popular to think this way in the philosophy and science of artificial intelligence, the idea that intelligence or consciousness can be manufactured in some way, for instance by software. Karl Marx thinks of capitalism proper emerging from the collective whirr of enough machines. When enough of them are connected and whirring away, pop! Out comes industrial capitalism. Ecological philosophers definitely think of Gaia like that, Gaia being the more or less personified whole that emerges from the functioning of Earth systems such as the carbon cycle or the nitrogen cycle, as the scientist James Lovelock first argued.[7]

But there's no reason to think that way. When you draw a set of things, the circle you draw around those things is always going to be bigger than that set, physically speaking. Otherwise it wouldn't be able to encompass them. But how a drawing looks isn't what it logically *means*. If everything exists in the same way, that means that wholes exist in the same way as their parts, which means that there are always more parts than there is a whole – which means that the whole is always less than the sum of its parts. It's childishly simple when you think about it this way. So how come it's so hard to accept?

It has to do with the legacy of monotheism. Even if we don't believe in God, even if we're agnostic, we keep retweeting monotheistic concepts. Or our concepts have a monotheistic form, despite what we think we believe. That kind of holism, which I'm going to start calling *explosive holism* (in which the whole is always bigger than the sum of its parts) is just like that. God is omnipresent and omniscient, so God must be way bigger than the sum of the parts of the universe that He created (assuming it's a *he*). Or think about that fiery early American sermon: we are all sinners in the hands of an angry God.[8] He's so high, you can't get over Him, He's so wide, you can't get around Him. My God is bigger than yours.

The idea of sinners in the hands of an angry God sums it up. We are small, and furthermore we're *ontologically* small: we don't matter as much as God does. Naturally His human stand-in on Earth, the King, matters a whole lot more than us too. Kings and gods emerged in early agricultural (Neolithic) society. When you settle down and start farming you get a picture of the static social space you're in (hence the concept of 'the state' – hunter gatherers wouldn't think of organizing things in this way). This social space seems obviously bigger than your little part of it, and there's a strict social hierarchy (that emerged, along with patriarchy, within a short time of the beginning of the Neolithic Age). And there's division of labour: the King is the king, you're the blacksmith, that guy over there is the sesame merchant. All together we make up a whole that seems so much 'bigger' than the sum of its parts. But this is just an aesthetic picture, a sort of massive compression, like a really low resolution jpeg, of the existing social structure with its monotheism and its King and its division of labour.

There are lots of things in our world that work according to some idea that wholes are less than their parts. For instance, in the USA, the tax code is such that if you are married, the two of you count as one and a half people. That means that when you are married you become three quarters of a person. There's some kind of deep psychological truth to this. Being connected means there's weirdly less of you because you are being open and less of your ego is in the way.

Perhaps any relationship is like this. Perhaps that's what we've been getting wrong about marriage. Maybe in the West we think that for things to exist they have to be constant. In the philosophical lingo this is called *the metaphysics of presence*. So we think that marriages must be permanent. We think that when they fail there must be something wrong. But if we are generous we realize that relationships are all different, and perhaps all relationships are finite. What if we added to this insight a way of distinguishing between *infinity* and *permanence*? A marriage might have infinite depth but still be impermanent. Think of a fractal shape. Mathematically speaking, if not in actual reality, it could have infinite parts. But you can hold it in your hand. Maybe this is what the poet Blake meant. Maybe it's not so mystical at all when he wrote: 'To see the world in a grain of sand . . . Hold infinity in the palm of your hand.'[9] Blake understood the pitfalls of agricultural-age religion, how oppressive it could be. In the same poem, he talks about the horribly broken state of England, which at the time was on a war footing, by analogizing it with how humans treat animals: 'A dog starv'd at his Master's Gate, / Predicts the ruin of the state.'[10] Doesn't this say something along the lines we are thinking? The way things are now, there doesn't seem to be quite enough room in our ideas about the state, let alone our house, for nonhumans. Yet we possess them anyway. And beyond that possession, they are part of our world, occupying our built spaces. So are the hibiscus flowers in my street, bursting up through the broken Houston concrete. We didn't invite them, but they're here anyway.

Perhaps that's what's wrong with most human-built space in what is called 'civilization', that it doesn't accommodate the beings who are already here, walking around as strays or bursting through the cracks in the concrete. These nonhumans are like uninvited guests. With human uninvited guests, we follow rules of hospitality, we welcome them in (unless they are hostile) and ensure they don't feel like their arrival is a disturbance, even if it is. But with nonhumans, what's the etiquette? Well, we are perhaps reaching the point where we might want to revisit our customs, our rules and modify them to include at least some nonhumans.

How things exist is mashed together with how they appear for other things. A tree isn't connected to the forest it's in just because it's measurably 'inside' the forest. The tree *has to do with* the forest. Being-part-of-a-forest is one of the ways in which it appears: it sucks nutrients from the forest floor; it communicates with the trees in its neighbourhood; it provides a home for squirrels. So if we're going to apply to the tree what we were just thinking about hammers and swapping hardened sausages for official hammers, then we would see that being-in-a-forest doesn't exhaust being a tree. That's just one thing a tree can be.

There are all kinds of things that logical sentences are that don't have to do with my brain having them (if brains do have them – we still don't really know). Yet they show up, these sentences, 'in' me to the extent that I am 'into' them. In exactly the same way, trees aren't just symptoms of forests. That's just one thing that they are. Things are entangled with interpretations of things, yet different from them.

Weather isn't just a symptom of climate. Rain can be an irritating cold sensation on the back of my shirt at 7.30 a.m. as I walk my son to school. Rain can be a wonderful bath for this light brown mourning dove on my balcony. Rain can be a refreshing drink. But rain is definitely caused by climate. This tree is definitely part of this forest. This kind of thought is definitely something that typifies me, Tim Morton, the thinker of that thought.

Things are much more mashed together than we like to think, and also much more distinct. The biosphere is made of its parts. But it's distinct from its parts. Which in turn means that its parts aren't reducible 'upwards' to the biosphere. Which in turn means that the way we have been thinking about things like biospheres is quite, quite wrong. Again, we call this way of thinking holism, and what holism means normally is that the whole is always greater than the sum of its parts. The parts are swallowed completely by the whole, like salt dissolving in water. This isn't how things really work – including salt solution. Think of a cello and a piccolo playing notes in a room. One is high-pitched and squeaky, the other is low and velvety. The two notes don't somehow blur into one another to form some amalgam that cancels out the distinctiveness of the piccolo and the cello. There isn't a piccocell or cellopicc or what have you that gets created out of the two. Yet the sounds relate, one to the other, and together they make a chord. The chord is distinct and has distinct effects. There's the metallic, breathy timbre of the piccolo, and the gritty, stringy woody timbre of the cello, like two ingredients in a cocktail. The cocktail tastes different from the whisky and bitters that make it up. But once the cocktail has been stirred, this doesn't mean that there is no whisky any more, that what whisky does is totally exhausted by the cocktail.

The whole isn't greater than the sum of its parts. In fact, the whole is *less* than the sum of its parts. This sounds so crazy that we are going to have to think it through a few times. But once you get it, then it really does seem like a much easier way of thinking. And it's a much nicer way of thinking – nicer to parts, which in our case, the ecological one, means nicer to polar bears and coral. The normal kind of holism is really a mechanism, though ecological thinking often dresses up this mechanism in nice green costumes, like a camouflaged soldier. The soldier still has a gun and could kill you. Mechanism with soft green bits is still mechanism. Mechanisms are things where the parts are replaceable. If your starter motor breaks, you can get a new one. The component itself doesn't matter. This is a very dangerous idea, ecologically speaking. Individual species don't matter. It's the good of the whole that's important. But if the whole and the parts are distinct in such a way that the whole doesn't totally swallow up and dissolve the parts, the parts matter a lot.

I think we have simply been passing on the normal form of holism without thinking too much. We do this because we are unconsciously reproducing good old agricultural-age theism that way. My God is bigger and badder than yours. If when things relate together we can forget them and concentrate just on the super-being, the network that the things create, we can ignore extinction. Something else will come along and work just as well as this lifeform that's going extinct. The biosphere will work just fine with jellyfish, which is the lifeform some people think will survive a massive amount of global warming. There are no clownfish, no coral reefs, no humpback whales, no sea sponges. Who cares? Life carries on. If that's what you call living, maybe I don't want to have anything to do with it.

That Earth-from-space image we now carry around in our heads should tell us something. That little blue marble, that 'we've got the world in our hands' feeling, is different from the coral and polar bears (and so on) that make it up. It's not a pink grapefruit. It's not a green crystal. It's a blue marble. It has specific intrinsic qualities. And, just like in the photo, it's *smaller* than its parts. I don't mean physically smaller. If you measure it, obviously it's much bigger: it's the whole of Earth. I mean that it's *ontologically* smaller. Ontologically means having to do with its being, not with exactly how it appears, with data that you can point to on a screen or measure with a ruler or touch with the tip of your tongue. Earth is one. A polar bear is one. There are lots of polar bears and coral reefs and parrots. There is one biosphere. Simple. The whole is less than the sum of its parts, because the whole is one, and the parts are many, and things exist in the same kind of way, if they exist at all. Let me explain.

This idea, that things exist in the same kind of way no matter what they are, is what some people call *flat ontology*. It's a little weird at first, but it takes the pressure off, I can tell you. And it gets rid of a lot of gnarly paradoxes that result from clinging to an idea that some things are less real than others because they can be reduced to other things, like their parts or the whole of which they are part, to take the items on the current agenda. If for example you think that the atoms of which you are made are more real than yourself as a person, then you will have to explain how your personhood arises from that atomic level, and that might be virtually impossible. And you now have the problem of explaining how the atoms arise. But most importantly, you have to justify why your idea of 'real' means that atoms are real in a 'better' way than medium-sized things such as horses or humans. Science never claims that atoms are more real than tomatoes – it restrains itself from ontological arrogance.

That's how things are related. They are related more like the notes coming out of a piccolo and a cello in a drawing room one sunny Sunday afternoon. They are distinct and the whole of which they are parts is distinct. Things aren't related as in some flavourless stew where the ingredients have totally dissolved. Modern physics is starting to say the same thing, which is encouraging. There are electrons and there are Higgs bosons that give the electrons mass. But the electrons aren't reducible to the Higgs bosons. Maybe there is a gigantic ocean of gravitons that give everything space. But the way this all works is that electrons are like little oceans within a much larger ocean of the Higgs field. These little oceans are distinct, they aren't protons, yet they are also part of the more general field, and so are protons. This is like saying that there is a specific orchestra playing specific Ludwig van Beethoven's specific Fifth Symphony, and within that there are specific instruments. The cellos are not the piccolos are not the trumpets are not the timpani . . .

and this isn't the Vienna Philharmonic Orchestra, it's the Royal Philharmonic, and this isn't the composer Gustav Mahler, it's Beethoven. There isn't this sludge called 'music' of which Beethoven's music is just a kind of specially shaped lump. The biosphere is not the sludge to which everything can be reduced. Thinking that way would be terrible. And very complicated. If that was how it was, how could oak trees be different from pandas, how could they ever emerge from grey sludge?

We have a problem with words such as *part* and *whole*, *specific* and *general*. In a way, a whole is really another kind of specific, not a generalization about specific things. This means that there is a weird gap between the whole and the parts, again, an *ontological* gap. You can sense this too in the weirdly small, fragile blue marble Earth pictures versus the gigantic trees you see around you while standing in a rainforest. There is a sudden perspective jump between these two images. You could imagine a camera zooming in and showing you the rainforest trees inside that blue marble. This is how a lot of environmentalist imagery actually works. But this is just an attempt to smooth over this ontological gap. Maybe it's better to think that there is always a sudden, quantum-like jump between different scales, because there is an ontological jump between a thing and its parts. If they were sentient, the atoms in a kettle of boiling water wouldn't be experiencing anything like the smooth flow of steam that comes out of the spout. Their electrons are jumping suddenly and randomly between lower and higher orbits. Maybe the smoothness aesthetic is on the wrong side. Maybe it has a side effect of making us feel that we are inside a gigantic machine, and the camera is like God, giving us the thrill of seeing all the way down into His world, being able to inspect and replace every component at will, because we can zoom in and out so smoothly: a feeling of omnipotence. And how's that idea been working out for us and other lifeforms so far?

The 'quantum jump' feeling is much more accurate, experientially. And this makes me think that it's because of a deep feature of reality, that it's jumpy because things are distinct and unique. You are in a plane, descending, descending, when suddenly, zoom – we are almost landing and there is a whole different feeling, probably a lot more adrenaline-fuelled and queasy, as we start to relate to the ground. You are fumbling around vaguely and suddenly, zoom, you appear to be married with kids. As David Byrne (whom we met a little earlier) asks, 'Well – how did I get here?'[11] That, the feeling of uncanny dislocation, as we learned in the previous chapter, isn't an optional extra. It's the only game in town.

The point being, *You already got here.* And this realization gives us a clue about what ecological action looks like, which in turn gives us a clue for evaluating ecological ethics and politics. That's how we'll be proceeding in the next chapter.

CHAPTER 3

Tuning

Let's think about the delivery mode of ecological advice – drive less, shop locally, save energy, all the usual 'shoulds' that we hear again and again. Either we are being preached to as individuals, being made to feel bad and encouraged to change our habits, so that maybe we will feel better, because we think others think of us differently – or we are being lectured at, made to feel powerless, because the thought of revolution or other big kinds of political change are very inspiring, but also bring up thoughts of how they might be resisted or constrained: the powers that be are too great, revolutions are always co-opted . . . Maybe they're just impossible on any scale that would matter. Sometimes I think, 'Really? I have to assemble a huge group of humans and start a revolution right now, *then* I can relate to polar bears?'

But awareness of the sensuous existence of other lifeforms doesn't have to involve big ideas or actions. How about just visiting your local garden centre to smell the plants?

Why this constant and very particular orientation to the future – what needs 'to be done' in order to start being ecological? It's a sort of gravity well that ecological thought about ethics and politics can get stuck in. You think *future* and you think *radically different from the present*. You think *I need to change my mindset, now, then I can really start making a difference*. You are thinking along the lines of agricultural religion, which is designed mostly to keep agricultural hierarchies in place. You are trying to get the right attitude towards some transcendent principle; in other words, you are operating within the language of good and evil, guilt and redemption. Agricultural religion (Judaism, Christianity, Hinduism and so on) is implicitly hierarchical: there's a top tier and a bottom one, and the very word *hierarchy* means *the rule of the priests*. By framing ecological action this way, you have been sucked into a gravity well, and it's not an especially ecological space down there. In many ways, it's not helping at all. For instance, there's really no reason to feel individual guilt: your individual actions are statistically meaningless.

We don't have to frame an ecological future as being radically different, at least not in quite that way. Now some of you may be tempted to close this book because you've already pegged me as a quietist who doesn't want to address the elephants in the room such as neoliberal capitalism. You'd be quite wrong. I'm talking about exactly *how* to address the elephants, considering that all forms of elephant address so far haven't worked out so well for planet Earth (and all the creatures, including humans, who live on it). There's nothing wrong with being a little bit hesitant and thoughtful and reflective. But anti-intellectualism is the favourite hobby of . . . the intellectual. At the end of ecology conferences, you so often hear someone saying, 'But what are we going to *do*?' And this has to do with guilt about sitting on chairs for a few days thinking and talking (and perhaps also with the sheer physical frustration of sitting on chairs for a few days).

I want to take an entirely different approach. I want to persuade you that you are *already* being ecological, and that expressing that in social space might not involve something radically, religiously different. Don't think this means that nothing changes, that you are just the same when you know about being ecological. It's rather hard to describe what happens, but something does happen. It's like someone slit your being with a very sharp and therefore imperceptible scalpel. You started bleeding everywhere. It's something like that.

A couple of years ago, I was being interviewed for a magazine. The interviewer was asking a lot of devil's advocate type questions, so many in fact that I started to think that they weren't devil's advocate questions at all. I started to think that he seriously didn't like the idea of acting ecologically. I wondered how I was going to convince him. Then I wondered whether convincing mode was the best way of addressing his stance. As I've just described, this mode might have some bugs in it, bugs from religious discourses that were originally set up in part to justify a massive firewall between humans and nonhumans (cattle over here, frogs over there, cats charmingly – or suspiciously, perhaps – in the boundary space between here and there). And ecological action is very evidently about not having such a firewall.

Then something occurred to me.

'Do you have a cat?' I asked.

'Yes,' he replied, perhaps somewhat taken aback by the oblique and simple question.

'Do you like to stroke her or him?'

'Oh, yes of course.'

'Well, so you're already relating to a nonhuman being for no particular reason. You're already being ecological.'

The journalist didn't like it. Conventional wisdom says that being ecological is a special, different mode of being, akin to becoming a monk or a nun. And the theory of action that fuels this special being also has a religious patina to it, in an antiquated way. Let's consider a different approach altogether.

It's going to take us a little while to get the hang of the 'no particular reason' part of the above statement. And it's going to take a while to determine exactly what 'relating to' means. Both have to do with a concept that I'm going to call *tuning*. I think we are already being ecological – we just aren't consciously aware of it. And those of us who *say* they're being ecological might be saying it in a mode that doesn't have anything in particular to do with coexisting nonviolently with nonhuman beings, which is roughly what I take ecological ethics and politics to mean. This nonviolence doesn't have to be as extreme as Jainism, perhaps. And perhaps it can't pretend to be perfect or pure. It's fraught with ambiguities, because sharks can eat you and viruses can kill you and it would be a good idea to protect our human selves from viruses and sharks. Furthermore, we can't determine in advance how wide the net of our concern should be, because we don't know everything about all lifeforms, and we don't know how they are all interrelated – and our actions cause further interrelations, tangling us even more. Nonviolence in this respect is uneasy and shifting.

Free Will is Overrated

We have, incidentally, made some ethical and political progress in the last couple of pages, though you may not have noticed. One thing that we just got clear is that it's possible to combine traditional environmental ethics and politics with animal rights ethics and politics. Though they seem like they might be naturally akin, some people regard joining up these two discourses as an impossible task, like squaring the circle. Environmentalism and ecological science is often about populations rather than individuals, and populations are considered very differently from individuals – in ways, animal rights critics might argue, that are insensitive to specific nonhuman beings: how they can be managed and controlled, for example. Animal rights talk, on the other hand, is frequently concerned with specific individual lifeforms – how they suffer, how they should be treated – even if there are many of them. But the seeming difference in focus between these two types of thinking may not be as distinct as it seems, and it has to do with something we've been exploring, namely our trouble with thinking wholes and parts. Let's consider the sharp distinction between what is considered to be an environment (or ecosystem) and a lifeform (individual animal).

We think, for example, that ecosystems (and populations of lifeforms, for that matter) are wholes with parts that relate to them mechanically, in the sense that they are replaceable. If there's something wrong with your engine, you replace a component and it's fixed. Science is ethically neutral but you can imagine using ecological science to justify a certain kind of unpleasant ethics. A lifeform goes extinct? Never mind, the whole will generate a new component to take its place. You can imagine that this doesn't work very well for the animal rights crowd.

But we are also going to need to have a little conversation about rights. If the choice is between mechanical wholes and separate individuals defined according to the normal manuals for defining such things, I don't want anything to do with either. They might actually be two halves of a torn whole, as one philosopher, Theodor Adorno, liked to put this sort of thing. The trouble is that rights and citizenship and subjecthood (and languages related to those concepts) have to do with possessing things. Individual rights are based on property rights, so that *being in possession of yourself* is one criterion for having them, for example. But if everything has rights, nothing can be property, so nothing can have rights. It's as simple as that. Scaled up to Earth magnitude, rights language doesn't work at all. The other problem is that to grant someone rights, you traditionally have to show that the someone is indeed a someone, in other words, that such a being has a self-concept. So the poor chimpanzee, to take an example from American law, has to wait around until enough humans are kind enough to condescend to grant it a self-concept. So far, such an approach has not been working out so well for the chimpanzee – or most other nonhuman creatures either.

This is why what Ecuador did in response to the oil corporation Chevron was so fascinating. Thirty thousand Ecuadorians living in the Amazon rainforest brought a $27 billion lawsuit against Chevron for drilling the Lago Agrio oil field, saturating the topsoil with viscous oil. From 2007 to 2008, Ecuador rewrote its constitution to allow for the 'rights of nature'.[1] This means that the nonhuman world has the right to exist and regenerate. If you think this is dangerously anthropomorphic, then too bad. The problem is that there is no other way for us as humans to include nonhumans within rights language than to bring them under the human umbrella under which we are sheltering. The difficulty is, many of the tools we have for making correct decisions are contaminated in advance with anthropocentric chemicals, as we will see in the following paragraphs.

The division between *act* and *behave*, which is based on a medieval Neoplatonic Christian doctrine of soul and body, structures how we distinguish between ourselves (the ones we allow to *act*) and nonhumans (the ones we only believe to be *behaving*, like puppets or androids). But are *we* Neoplatonic Christian souls? Isn't being a person a little bit about being paranoid that you might *not* be a person? Can you get rid of the ambiguity without tearing something?

There is an additional issue. We observe some emotions in nonhumans such as elephants, but we are less willing to let elephants feel emotions that seem less 'useful' to us. We can let elephants be hungry when they look hungry, but we have trouble allowing that they are happy when they look happy.[2] That, for some reason, would be anthropomorphic, and many environmentalist thinkers are concerned not to be, although I've argued that it's impossible, since even if you intend not to be, there you are, a human, relating in whatever human way you are relating to whatever other lifeform. It's interesting that we think that sheer survival (hence hunger) is more 'real' than some kind of quality of existing (such as being happy). It says a lot about us that just surviving, being hungry, are supposedly 'real', aka nothing to do with being human in particular – what does that say about us and what does it in fact do to us ourselves, let alone the elephants? Ecological catastrophe has been wrought in the name of this survival, sheer existing without heed to any *quality* of existing. Objectively, in terms of how we have acted it out, this default utilitarianism has been very harmful to *us*, let alone other lifeforms. That says it all, doesn't it? It's like that language about the bottom line. We may feel bad about workers suffering, but profits must be maintained, corporations must go on existing for the sake of existing. These two types of thought – about survival and bottom lines – are synonymous.

The environmental approach could be described as taking care of the whole at the expense of individuals, while the animal rights approach could be described as taking care of individuals at the expense of the whole. We seem to be at an impasse. But notice a feature of the two approaches. The 'take care of the whole at the expense of the individuals' and the 'take care of the individuals at the expense of the whole' approaches do share something. They are trying to give you a good reason to care about nonhumans. But what if *having a good reason to care* was precisely a large part of the problem? Getting a bit more granular, animal rights and environmentalism give reasons that are reductionist. Reductionism doesn't necessarily mean that large things are made of small things that are more real than large things. Sometimes we can reduce small things to large things. The environmentalist approach defines wholes as more real than (and so more important than) their parts, or they describe parts as more real than (and so more important than) wholes.

We can start to break through this difficult impasse by noting that what is called *environment* is just lifeforms and their extended genomic expressions: think of spider's webs and beaver's dams. When you think this way, you are already thinking about wholes and parts in a different way.

And when you think of things like *that*, there's really no difference between thinking about what is called an ecosystem and what is called a single lifeform. Problem solved.

Thinking about wholes and parts in this way is a key component of good old-fashioned art appreciation theory. A work of art is a whole, and this whole contains many parts – the materials out of which it's made being just one of them. We could include the interpretive horizons of the art's consumers, for example, and the contexts in which the art materials were assembled – a highly explosive concept, as we saw earlier. In this way it's obvious that there are so many more parts than there is whole. In an age of ecological awareness there is no one scale to rule them all. This means that art and art appreciation won't stay still, in the way that a lot of art theory (for instance in Kant) wants. And in the absence of a single authoritative (anthropocentric) standard of taste with which to judge art, how we regard it is also about how wholes are always less than the sums of their parts. A work of art is like a transparent bag full of eyes, and each eye is also a transparent bag full of eyes. There is something inherently weird, even disgusting, about beauty itself, and this weirdness gets mixed back in when we consider things in an ecological way. This is because beauty just happens, without our ego cooking it up. The experience of beauty itself is an entity that isn't 'me'. This means that the experience has an intrinsic weirdness to it. This is why other people's taste might come across as bizarre or kitschy.

The truth is the choice to be able to care or not care is always an illusion anyway. You are always in care space, always in truthiness (as in the previous chapter). If you say 'I don't care about this issue,' it means that you care about this issue enough to say that. Often, in the real world, saying you don't care much about someone or something means you might be hiding that you care very much indeed.

Consider the phenomenon of 'single source recycling' where you don't have to sort stuff into plastics, cardboard, organic waste and so on: your bin cares about the recycling, so you don't have to. Some environmentalists have objected to it, visiting houses in my home town of Houston, Texas, for example, and persuading people to sign petitions. But why? Why the search for hypocrisies in the new process? Because it eliminates the idea of free will, and the performance of 'look at me I'm doing good'. The idea that we're outside the world looking in, deciding from a menu which choice to make, is precisely the dangerous illusion.

When you play a game such as cricket or baseball, the ball arrives at your bat within a few milliseconds. That's faster than your brain. You can practise and practise so that you can hit that ball when it arrives. That sounds elementary. But if you think about the fact that the ball is still faster than your brain, what on Earth is happening? Whatever is happening is a direct refutation of the Neoplatonic Christian idea we are still retweeting, that we have a mind-like or soul-like thing that is somehow inside us like a gas in a bottle, totally different from that bottle in some way, and that it is a sort of puppet master pulling the strings. You think you are about to hit that ball, but you have already hit it. Free will, as I keep saying, is overrated.

But it's even more strange and interesting. Consider an actual scenario. The fastest cup-stacker on Earth (a young boy) competed with David Eagleman, a neuroscientist, on his show, *The Brain*, which ran on PBS in America in 2015. They are wired up to brain scanners. The neuroscientist's brain is working overtime and he loses. The boy's brain is hardly working at all.[3] It's as if he is a zombie. He isn't intending to stack the cups and there isn't a puppet master inside his head pulling the strings. Something else is happening. His ability to stack the cups is *all* in his 'body'. Is the brain more like some kind of starter, which gets things going, then sits back? Well, we've just refuted that – the feeling of having made a decision might arrive slightly *after* you've made it, whatever it is. So the brain isn't even that, some kind of prime mover of a mechanism that keeps going once you've pressed a button. It looks as if what we're observing is neither mechanical (the latter option) nor orchestral (the former one). Some boss doesn't start the machine, and some conductor doesn't need to 'intend' everything all the time – as any concert musician will tell you (my father, for example), the conductor is never actually driving the music like that anyway.

Both these models have to do with a myth. The myth is that for something to exist, it must be constantly present: the metaphysics of

THIS IS

NOT HERE

presence. The soul-and-body, 'conductor' model seems up to date because it has to do with management, ownership and all kinds of things associated with the notion of private property that influence a lot of what we do on this Earth. But this turns out, as we have seen, to be a retweet of a Neoplatonic Christian concept.

Furthermore, the 'on switch' model of action depends on a mechanical theory of causation that requires some kind of god-like being at the start of the causal chain, to get the ball rolling. After that, the ball hits the next ball in a mechanical way. So the mechanical theory is really just a variant or upgrade of the 'conductor' one. And this is therefore merely a modification of our Neoplatonic retweet: the soul is the driver, the body is the chariot . . .

Let's make a new word: *alreadiness*. This word is going to come in very handy, because now I don't have to resort to a suggestive but rather clunky phrase from one of my favourite philosophical regions: deconstruction. This would be the famous *always-already* employed by Heidegger and then by Jacques Derrida, the inheritor of Heidegger's approach, which he called *Destruktion* ('de-structuring'), and which Derrida calls deconstruction.

Alreadiness hints at our tuning to something else, which is a dance in which that something else is also, *already*, tuning to us. Indeed, there are some experiences in which it simply can't be said which attunement takes priority; which comes first, logically and chronologically. One of these is the common experience of beauty. We can learn a lot from it: let's go.

You are Being Tuned

We could talk about our current historical phase in many ways: entering an ecological era, learning how to cope with global warming, and so on. But what all these labels have in common is *transitioning to caring about nonhumans in a more conscious way*. This talk is about that, and as you'll see it's a lot stranger than it sounds.

In November 2015 I participated in *Ice Watch*, Icelandic-Danish artist Olafur Eliasson's installation outside the Panthéon in Paris. *Ice Watch* was designed to be seen by the delegates representing the nations of Earth in the COP21 negotiations, otherwise known as the global warming summit, which was held over thirteen days. Eliasson and I recorded a public dialogue about it in Copenhagen about one week before *Ice Watch* was installed, at the CPH:DOX film festival. One thousand people attended, eager to hear about ecology and art.

Ice Watch consisted of something like eighty tons of ice harvested from Greenland and shipped intact to Paris, where it was installed in twelve gigantic chunks, in a circle. From above, it readily resembled the little bars that stand for hours on a wristwatch. The chunks of ice were large enough to climb on to and sit in, or even lie in, and as there was no barrier protecting them, this is just exactly one of the things that people did. Part of the project was documentation of all the different ways in which you could access the ice. You could walk past it. You could ignore it. You could touch it. You could reach out towards it. You could talk about it. You could give a conference paper about it at a conference called *Façonner l'avenir*. You could sleep in it. This was especially easy once the sun had melted the ice enough for it to form smooth pockets and contours.

Part of the point of *Ice Watch* was an obvious visual gag: look, ice is melting and time is running out. But that was just the hook. What actually happened was much more interesting, and in a way that seriously stretched or went beyond prefabricated concepts, in a friendly and simple, yet deep way. Watches are things that humans read. But they are also things that flies land on, things that lizards ignore, things that the sun glints off. Dust settles on the glass shell of the front of the watch. A dust mite traverses the gigantic overpasses and caves on the underside of the watch between the watch and my wrist. And let's return to something I just said about *Ice Watch*: the sun melts it. The sun is also accessing the ice. The pavement is also accessing the ice. The climate of Paris is also accessing the ice.

And the ice was accessing us. It seemed to send out waves of cold, or suck our heat, whichever way around. This kind of access was how Eliasson was thinking about it – the encounter with *Ice Watch* is in a way a dialogue with ice blocks, not a one-way human conversation in a mirror that happens to be made of ice. We've been having that kind of conversation with nonhuman things for thousands of years. It's exactly the reason we are in this mess called global warming. And the climate factoids we hear on the news are echoed by much of the art that tries to address global warming and extinction. For example, several artists have compiled massive lists of lifeforms that are going extinct. But the risk here is of becoming just like those factoids: just a huge data dump. Art is important to understanding our relationship to nonhumans, to grasping an object-oriented ontological sense of our existence. Art fails in this regard when it tries to mimic the transmission of sheer quantities of data; it's not artful enough. This isn't just a matter of effective persuasion. As a matter of fact, that's the trouble with ecological data art. The aesthetic experience isn't really about data – it's about data-*ness*, the qualities we experience when we apprehend something. (As I mentioned earlier, data just means 'what is given', and isn't only about numbers and pie charts.) The aesthetic experience is about *solidarity* with what is given. It's a solidarity, a feeling of alreadiness, for no reason in particular, with no agenda in particular – like evolution, like the biosphere. There is no good reason to distinguish between nonhumans that are 'natural' and ones that are 'artificial', by which we mean made by humans. It just becomes too difficult to sustain such distinctions. Since, therefore, an artwork is itself a nonhuman being, this solidarity in the artistic realm is already solidarity with nonhumans, whether or not art is explicitly ecological. Ecologically explicit art is simply art that brings this solidarity with the nonhuman to the foreground.

Eliasson wanted to do something that was logically prior to collecting data, let alone spreading it around. To collect data, you have to be receptive. You need the right kind of data-gathering devices for your project. You need to care. A global warming scientist needs to care enough about global warming for her to set up the experiments that find out about it in the first place. In the beauty experience, there is some kind of mind-meld-like thing that takes place, where I can't tell whether it's me or the artwork that is causing the beauty experience: if I try to reduce it to the artwork or to me, I pretty much ruin it. This means, argues Kant, that the beauty experience is like the operating system on top of which all kinds of cool political apps are sitting, apps such as democracy. Nonviolently coexisting with a being that isn't you is a pretty good basis for that.

Since the being that isn't you is artwork, and so not necessarily human, or conscious, or sentient, or for that matter alive, we're talking about the possibility of being able to expand democracy, from within Kantian theory itself, to include nonhumans. Which is a pretty scary thought for some people – Kant himself, for example, which is one reason why he is so careful to police the magic ingredient, the beauty experience, that actually makes the rest of his philosophy work (like Heidegger, he pulls back on his own thought, not carrying it through to its potentially radical conclusions). Instead, he sort of introduces a little tiny drop of it to flavour the anthropocentric – and pretty much bourgeois – soup – too much and the soup is ruined; it ceases to nourish anthropocentric patriarchy. It's funny that the way to undermine Kant, as with Heidegger, is to take him more seriously than he takes himself, a tactic I've definitely inherited from deconstruction. And you do it by increasing the amount of the very ingredient that makes the soup so tasty.

When you encounter the beauty experience, it's not about anything in particular. If it really was a bowl of soup, you might want to eat it. Then you'd know what the thing was about: it was about future you, with a nice full belly. In a way, you would know the future of this entity, this object, this bowl of soup. But because beauty soup isn't for eating – because it's just this weird slightly telepathic mind meld between me and something that isn't me – you don't know the future. There is a strange not-yet quality built into how you access the thing you are finding beautiful. And because, from my point of view, beauty is sort of like having data, but the data isn't pointing at anything but itself – I'm just experiencing the givenness of data, of what is given. I'm experiencing the way data doesn't quite point directly at things. That's why you need scientists, right? They figure out patterns in data that hint at things. That's why science is statistical. That's why the sentence *humans are causing global warming* is actually not at all like *God created Earth in seven days*. You don't need to believe it in a firm sense. You can just accept it as pretty much true. You can be 98 per cent correct, and that's better than threatening me with torture unless I admit that you're completely right, because there's no other way for you to *be* right than to hit me until I agree.

I'm also experiencing something magic and mysterious about myself when I have that beauty experience. The ice is a sort of Pandora's box with an infinity within it. And so am I. It's that mouthfeel again. I'm experiencing the texture of cognitive or emotional or whatever phenomena. I'm experiencing *thinkfeel*, or better, since I can't tell whether it's about thinking or feeling but I know it's real and it's happening, it's *truthfeel* that I'm experiencing. It's as if I could magically see around the corners of myself to the part of me that's having the thoughts, because when I try normally, I just find another thought. I can't see all of my phenomenological style, how

I manifest in a complete way, all at once – that total happening called 'me' is only accessible in slivers. Some people call this thing that keeps disappearing around the corner *consciousness*, Kant calls it the transcendental subject, but as we've seen, there's no particular reason to hold on to these concepts.

I magically see the unseeable aspects of a thing, including the thing called Tim Morton. I grasp the ungraspability of a thing. Which is another way of saying, I see the future, not the predictable one, but the unpredictable one. I see the possibility of having a future at all: I see *futurality*.[4]

And in the case of the *Ice Watch* hunks of ice outside the Panthéon in Paris, Eliasson set this up so that you could see this future isn't a container for the ice block. It's coming directly out of the ice block itself – the ice block is creating the future. The ice really is a watch. And not a watch being set by humans. Or even better, it's a certain kind of time structure – it is a temporality structure. It allows you x and y and z kinds of past and future. This is the paradox. Futurality isn't some grey mist that is the same for a block of ice as it is for an excited proton underneath Geneva. Different objects, different futuralities. Unspeakableness or ungraspability can come in all kinds of flavours. It only sounds paradoxical because we're used to time and space being box-like containers in which things are sitting, where we place and try to contain them (no matter whether this effort is an illusion or not), whereas for Kant, and those who come after him, time is something posited, it's part of aesthetic experience, it's in front of things, ontologically, not an ocean in which they are floating, but a sort of liquid that pours out of a thing.

So we have to be careful what we humans design, because we are *literally* designing the future, and that future isn't in our idea of the thing, how we think it will be used and so on – that's just our access mode. The future emerges *directly from the objects we design*. Right now, many, many objects on Earth are designed according to a one-size-fits-all, very old, way past its sell-by-date temporality template. It's one we have inherited from Neolithic agriculture, that's how ancient it is. And it's the one that has given rise to industry with its fossil fuels and therefore to global warming and mass extinction. So designers should be careful what they design. Maybe they need to think at least on a number of different temporal scales when they design something. A plastic bag isn't just for humans. It's for seagulls to choke on, and now we can see that thanks to photographers such as Chris Jordan who photographs beings who get caught in the Pacific Garbage Vortex. A Styrofoam cup isn't just for coffee, it's for slowly being digested by soil bacteria for five hundred years. A nuclear device isn't just for your enemy. It's for beings 24,000 years from now. This Diet Coke isn't just for me. It's for my teeth and my stomach bacteria, and the latter may get slaughtered by the acids in there. This is why I created the concept of the *hyperobject* in my book *The Ecological Thought*. A hyperobject is a thing so vast in both temporal and spatial terms that we can only see slices of it at a time; hyperobjects come in and out of phase with human time; they end up 'contaminating' everything, if we find ourselves inside them (I call this phenomenon *viscosity*). Imagine *all the plastic bags in existence at all*: all of them, all that will ever exist, everywhere. This heap of plastic bags is a hyperobject: it's an entity that is massively distributed in space and time in such a way that you obviously can only access small slices of it *at a time*, and in such a way that obviously transcends merely human access modes and scales.

Time Flows from Things

Everything emits time, not just humans. So when we talk about sustainability, what we're talking about mostly is maintaining some kind of human-scaled temporality frame, and this is necessarily at the expense of those other beings, and it's very likely we didn't factor them in at all. What exactly are we sustaining, if not the one-size-fits-all agricultural temporality pipe that has sucked all lifeforms into it like a vacuum cleaner, pretty much, over its 12,500-year run? And in the end, which means already, designing stuff according to that template is going to damage humans as well, in a very obvious way, because of the unavoidable interconnectedness of everything we know and understand, and even everything we can't know or see, too. When the Nazi propagandist Joseph Goebbels heard the word culture, he reached for his gun. When I hear the word *sustainability*, I reach for my sunscreen.

Everything we've been exploring in the last few pages occurs to you as ethical and political fallout from the Kantian beauty experience; as wonderfully open-ended, because the kind of futurality a piece of artwork opens up is unconditional: in other words, it doesn't have a rate at which it decays to nothing. You don't ever exhaust the meaning of a poem or a painting or a piece of music, and this is another way of saying that the artwork is a sort of gate through which you can glimpse the unconditioned futurality that is a possibility condition for predictable futures. Art is maybe one tiny corner in our highly (too highly) consciously designed – and way too utilitarian – social space where we allow things to do that to us. What would it look like if we allowed more and more things to have some kind of power over us?

This isn't quite the same thing as saying, along with the socialist William Morris, that functional things should be beautiful. That's because, on this view, things are just lumps without some nice decoration. But we're saying that there are no lumps. There are blocks of ice, humans, sunlight, the Panthéon, polar bears. The goal is not to take existing things such as sofas and houses and make them pretty in a way that working-class people can afford (for example). That kind of thing suffers from the same syndrome as sustainability: it's anthropocentrically scaled.

Likewise we can't do what we take to be the opposite, which is saying art is beautifully useless and if you can't appreciate it, that's your problem. Again, you are simply allowing its existing function for humans now – aka anthropocentric functioning – to be default. Art is a place where we get to see what it means to be human or whatever, which is why what I do is called humanities. But this isn't enough. One way this becomes obvious is when writing grant proposals that sound like pleading. Please, please don't hurt me, Mr Funding Source, I'm a sort of educated PR guy who is going to decorate this boring cupcake of scientism with these nice human-flavoured meaning-candies.

Realizing that there are lots of different temporality

formats is basically what ecological awareness is. It's equivalent to acknowledging in a deep way the existence of beings that aren't you, with whom you coexist. Once you've done that, you can't un-acknowledge it. There's no going back.

Enchantment: Causality as Magic

So far I haven't transgressed vanilla, basic Kantian Kant very much. Well, maybe the last bit. But now I'm going to push up some faders on the Kantian mixing desk that will add some more of the chilli flavours that he only allows in tiny droplets. Let's return to our poor grant applicant and indeed to our Arts and Crafts people, starting with William Morris. What is their language blocking? It's blocking the fact that art isn't just decoration. It's causal. *It does something to you.* The Platonists were right: art has an inherently disturbing (in a nice or not so nice way) effect, an effect that you don't intend and can therefore strictly be called *demonic*, in the sense that demons are the messengers of the gods: it's a message from somewhere else. Platonists accurately see the power of art, which is why some of them (such as Plato himself) want it to be banned or very heavily censored. An artwork does something to you, so if you think that only lifeforms can do things to you, this is a weird and challenging fact. If you think on top of this that only humans are empowered with the magical ability to impose meaning and temporality on things, then you are in for a bigger shock, because as I've argued, art emits time, which tells you something about how everything emits time. It's designing your future as much as you're designing its.

Kant only wants you to hear about 10 per cent of that, but it's a very important ingredient of the overall mix that you can't do without. But according to Kant, if you hear more than that, you are in danger of being *charmed* or *enchanted*, rather than experiencing beauty, and that, in his book, is not OK. It's OK to be wordlessly smitten by something, as long as you don't actually fall in love with it and ask it out on a date, or even worse, allow it to ask you out. He acknowledges that there is a mind meld, but only up to a point, and it really does have to do with how you're a human being imposing reality on things. So really, for Kant, the experience is coming from you, not the artwork. Mystery solved. Disenchantment in effect. We can relax. Kant didn't turn into Yoda. Which was on the cards, because he was fascinated with the paranormal (maybe in the same way homophobes are fascinated with homosexuality). He himself was entranced, but resented it or feared it. So while Kant had to allow the idea into his theory – mind melding with a nonhuman being is how the thing actually works – he did it in a contained way, not in a way that you'd notice, like a tiny subliminal droplet of Yodaness; a base to the soup whose ingredients you experience even if you don't know what they are.

By Yoda-ness I mean the actual Force, the one that eighteenth-century German physician Franz Anton Mesmer talked about, and which fascinated Kant: a sort of animal magnetism, a Force, argued Mesmer, was generated by lifeforms; it surrounds and penetrates them – it is like when Darth Vader makes a gripping movement with his hand, and not unlike how they used to mesmerize people with hand gestures, causing someone to believe they had been strangled – without touching them. Animal magnetism is to all intents and purposes identical with the Force of *Star Wars* fame; it is, as Obi Wan Kenobi observes, an 'energy field' that 'surrounds' and 'penetrates' us, and we can interact with it, with healing and destructive consequences.[5]

That's the problem with art, isn't it? It sucks you in, whether or not it's telling the truth, it's so truthy, it's not right or wrong but still it's giving off this incredible truth vibe, it's pulling me into its tractor beam, in a moment it might say, 'I find your lack of faith disturbing, Tim,' and strangle me, at a distance. Art is telepathic – it's spooky action at a distance, which is also what Einstein didn't like about quantum theory. It makes things happen without needing to touch things. But art is also profoundly ambiguous: we can't tell whether it's telling the truth or lying. Ambiguous *and* powerful at the same time for the same reasons.

Interlocked in the beauty experience, I might dissolve. The art thing might fit me so perfectly that I disappear. Turned up to 11, this My Bloody Valentine music will actually kill me. But I can't tear myself away from it. Resonating perfectly with the physical structure of this glass, an opera singer's voice causes it to explode. Maybe the beauty experience is like a little death warning light that goes off in my experiential space. Maybe beauty is death, in a way, just like the decadent aesthetes used to say. It's a reminder that things are fragile, because when one thing envelops another thing, that other thing might be overwhelmed or destroyed. Maybe when Oscar Wilde said, on his deathbed, 'This wallpaper and I are fighting a duel to the death; either it goes or I do,' he was telling the literal truth, and it only sounded like a joke because of our prejudices: the idea, for example, that appearances are superficial, while essences are fundamentally beyond appearance. The colour yellow shouldn't matter that much, we think. By the way, the wallpaper won.

So when I experience beauty, I am coexisting with at least one thing that isn't me, and doesn't have to be conscious or alive, in a noncoercive way, in which the possibility of death is vivid yet diluted and suspended. We coexist; we are in solidarity. I'm haunted, charmed, enchanted, under a spell, things could get out of control, but they won't, at least for now. The present moment collapses and I'm left with an uncertain, spectral futurality that is exactly what this chunk of ice happens to be. How it looks, how it feels, where it is sitting, its mass, its shape – all that, which we could call appearance, is the past. The ice chunk is a sort of train station in which past and future are sliding past one another, not touching, and what I mistakenly call present is a kind of relative motion between the two sliding trains of past and future. I call it *nowness* to differentiate it from a reified atomic 'present' that actually I don't think truly exists. A thing is exactly how the cookie crumbled, and how the cookie might crumble some more, and I get to coexist in this slightly sad, melancholic space where the crumbling happened, and where an uncertain future opens out. All cookies crumble, you know. That's why they can be cookies. Things are inherently fragile, they all contain a fatal flaw that allows them to exist, because they are always exactly what they are, yet never as they appear. They transcend all access modes but they are unique and distinct. The rift between being and appearing is ontological, in other words you can't point to it, it's intrinsic to a thing and it's why cookies can crumble. Even black holes evaporate.

And because it's not anthropocentrically scaled in particular, or ego-scaled in particular, when you have a conversation between beauty and disgust or ugliness, you can't delete it. It is a conversation between objects and abjection, which is a technical term some thinkers use to describe the functions of the body and the body's relation with its symbionts, against which the traditional Western human subject has learned to distinguish him- or herself. The more we know about objects from the OOO point of view, the more we realize that we can't cleanse them of their 'abject' qualities, because they aren't pristine, pure things, but pockmarked and pitted and oozing with all kinds of inconsistencies and anomalies – just like human beings. And because you're in truth space, you are having a conversation with actuality, even though it might not be your actuality or a human actuality. The artwork can't simply be a representation. The thing might have designs on you, to use the common English phrase. You feel this in the gravitational pull, the telepathic charm of the thing. And because of that, you are also having a conversation between having a purpose or a function and being beyond purpose or function, because a thing's function or purpose doesn't exhaust it. It just might not be your design or function or a human purpose in particular. Which is the same thing as saying you are having a conversation with utilitarianism, which is saying that you are having a conversation about happiness – whose happiness, and what kinds of happiness? Which means that you are having a conversation with what is probably something you think of as an inanimate object, like a block of ice, which means you are allowing yourself to be in a telepathic mind meld with something that stands for the worst possible fate of a human subject, being turned into an object. And because the truth space is truthy, not obviously truth as such, but saturated with truthiness data, you don't know whether it's true or not – the artwork is a lie that is telling the truth, or maybe it's a kind of truth that is lying. You are being telepathically seduced by a being that might be lying.

Actually *beings*, plural, so it's much, much worse, or better. Because there are so many more parts in the artwork than there is the whole of it, by definition, and by definition you're not allowed to discriminate either way – parts are more important than wholes, or vice versa, or one part is more important than the others – because that's finding a definite purpose, and the experience doesn't have that going for it. That would ruin it. This is due to that feature of OOO theory which we've already met (and which I'm advocating here), in which there is always a multiplicity of parts that exceed the whole, rather than the whole swallowing the parts perfectly. An artwork is *subscended* by its parts. We've already been exploring the concept behind this term quite a bit. Recall what I've been arguing already: that wholes are bursting with their parts; in a basic but strange-seeming way, *wholes are less than the sum of their parts*.

Those parts are also little temporality structures, little train stations within train stations, multiple tractor beams pulling you in, multiple hypnotists. Possibly an infinite regress of them – you can't check. Because you know you can't reduce that blob of paint to something it isn't, such as its parts (like little crystals or whatever, or brushwork), you can't delete its causal pressure on you. You decide that free will is most definitely overrated and we are going to need some kind of chemical to coexist other than rights and subjecthood and citizenship and free will. Infinity portals beckoning. Maximum aesthetic suction and repulsion, like a horror movie superimposed on a porno. And you still can't stop looking. It's not transcendent beauty, but it's still beauty. Which is another way of saying, it's not your bourgeois subject's best friend, more like an anarchic revolutionary army of little squirming pieces crawling around and within that seemingly rigid and singular piece of cheese.

Kitsch is the subscendent part of beauty, ghosting official anthropocentrically scaled forms of beauty like a spectre. In a way, kitsch or disgust is the X-power (as in the X-Men) of beauty itself. Without it, beauty can't evolve.

You have gone crazy, maybe.

All those things that Kant tries to edit out are back in, without deleting the beauty experience as such. In fact, they are deeply how it works, what it can't do without.

No Design is Perfect

This isn't the normal utopian or left way of critiquing theories about our relationship to arts, or aesthetics. The normal way is to say that art is only a construct and doesn't really exist – for example, it's just a bourgeois human ideology reproduction mode based on inherited ideas of taste. But what I'm saying is that art is actually a tiny but still recognizable fragment of the kind of larger world, the mostly nonhuman world of influences and designs that go beyond us and violate our idea of who 'owns' what and who is running the show, such that causality seems to have something animistic or paranormal about it. It's not a glue that falsely fixes bourgeois dichotomies such as subject and object. I'm talking about a substance that is a dangerous toxin to anthropocentrism and mechanical causality theories and the law of noncontradiction and default utilitarianism. The law of noncontradiction, for example, is an important lynchpin of Western philosophy, but it's never been proved, only stated, first by Aristotle in section Gamma of the *Metaphysics*. It is easy to violate and also easy to draw up logical rules that allow for some things to be contradictory. Since ecological entities are contradictory by definition (they are made of all kinds of things that aren't them, they have vague fuzzy boundaries . . .), we had better permit ourselves to violate this supposed law, at least a bit.[6]

Art only half works as a human-scaled bourgeois ideology reproduction device if you put just a tiny drop of it into the soup, and don't examine it too carefully or treat it as decoration. If you did, you would see all the subscendent little microbes squiggling around inside it, all of them trying to hypnotize you.

And this encounter with art tells us something about the encounter with any designed thing at all. Which is why you can sleep in an ice sculpture, which people were seen to do in Olafur Eliasson's *Ice Watch*. Or why tourists can take selfies in front of it, and there's nothing you can do about it. A thing is bursting with parts and scales and temporalities and sexualities, so a thing is never totally keyed to our taste or to a standard of good taste, but somehow that doesn't mean it's always definitely only ugly or that beauty and ugliness are false categories. It means that beauty is wild, spectral, haunting, irreducible, uncanny. And causal. Which means that the art versus craft or art versus design distinction breaks down, while leaving the difference between what a thing is for and its openness, its futurality, intact. 'Beautiful' is often said to be the opposite of 'useful'. It's held to be an unnecessary inconvenience, which is why so much of the modern world is so ugly. But beauty and usefulness *and* uselessness can't be separated at all. So every decision is a political one. Allowing a watch to be a landing strip for a fly. Allowing a plastic bag to be a bird murderer. Allowing a painting only to be seen by people who can afford the entrance fee. Living in a building designed to shunt dirty air somewhere else, where now we realize that *somewhere else* just means *nowhere else, because it's on the same planet.*

And irritatingly or wonderfully, this inbetween-ness means you can never have the perfect design. Because interconnectedness doesn't mean that there is an obvious whole that obviously transcends its parts and is bigger and badder and better than the parts, and the parts are just components in the machine of the whole. A political system is also a designed thing, so this definitely affects what kinds of future politics we want. Including bunny rabbits means excluding diseases fatal to bunny rabbits. I mean this quite literally. Because of interdependence, when you take care of one entity or group of entities, another one (or more) is left out. Biocentric ecological philosophy is quite wrong to claim that the AIDS virus has the same right to exist as an AIDS patient. You have to choose. Obviously I'm going to choose the AIDS patient.

And because of the gap between being and appearing, to be a thing *at all* is to be deeply flawed; in order to exist at all you have to have an intrinsic invisible crack running all the way through you. So a network of things can't be perfect, and a thing on its own can't be perfect. You can't seal off the futurality, you can't stop time leaking out of things and misbehaving, you can't reach the end of history, which now includes the history told by trees and geological layers and weather patterns. You just have to design your street knowing that, at some point, frogs are going to be crossing over it. At some point, it will be part of a geological stratum. At some point, a glint of light will reflect off a small puddle of water, blinding a driver and killing a pedestrian. At some point . . . The road is open, yet it's just this exact road, this black tarmac thing with white stripes on it.

And this tells us something about design. Humans can do it. But nonhumans also do it, all the time. Think about evolution. It's design without a designer. And in a larger sense, nothing is un-designed. There is no such thing as unformatted matter, waiting for someone to stamp a form on it. That's an ecologically dangerous fantasy of so-called Western civilization. In truth, anything at all is in part a story about what happened to it. My face has been designed by acne. A glass has been designed by glass blowers and cutters. A black hole has been designed by gravitational forces in a gigantic star. And in particular, things are definitely not unformatted surfaces that can only be formatted by human shaping or desire projection.

So the question is, with whom or what are we going to team up, and what kinds of affordances are we going to allow future beings, and how do we allow the spooky suspension of violence, the possibly infinite vortices of pleasures and pains with us and without us, like an eye that turns out to be a bagful of hypnotic eyes, to happen without collapsing it so fast? Because we've been in the collapsing business for quite a long time, we're really good at it, and now it's not just killing the bees, it's even killing *us*. So instead, I let the subscendent beauty of the artwork hold me in its infinite tractor beams, like a bagful of hypnotic eyes. What to do with these uninvited guests? Let them stick around, I guess.

Actual beauty has a 'Christmas tree effect': there is a greasy pathway towards kitsch, in which we become aware of beauty's 'disgust fringe' – there's more subscendent beauty than normalized beauty can cope with. And when I talk about art, it is not just as a metaphor for us to understand the quality of existence. The subscendent nature of art means that ecological art that calls itself as such can't be about Sierra Club-style uplifting poster-type grandeur. It must include ugliness and disgust, and haunting weirdness, and a sense of unreality as much as of reality.

And in turn, ecological awareness can't just be pure and pristine and holy. Why can't there be an ecology for the rest of us? For those of us who don't want to go out camping in the fresh air, but would rather pull the covers over our heads and listen to weird goth music all morning? When can we start laughing, not just in a hale and hearty way, but with irony, a sense of the ridiculous, an excessive feeling of joy? What would an ecological joke sound like?

The Manner of Attunement

I'm going to start this section with a quotation from psychoanalyst Jacques Lacan: *Les non-dupes errent*. It's a pun on his own phrase '*le nom du père*' (the name of the father) or '*le non du père*' (the *no* of the father).[7] Both of these phrases relate to how we understand the symbolic order under which we live – how we internalize power structures such as patriarchy and give language to systems of power. So what Lacan means when he turns these phrases on their head is that if you think you've got it right, that you can see through everything, that's when you've got it most wrong. Of course, the funny thing about the sentence is that it's subject to its own truth.

Since a thing cannot be known directly or totally, one can only attune to it, with greater or lesser degrees of intimacy. Nor is this attunement a 'merely' aesthetic approach to a basically blank extensional substance. Since appearance can't be peeled decisively from the reality of a thing, attunement is a living, dynamic relation with another being – it doesn't stop.

The ecological space of attunement is a space of *veering*, because in such a space, rigid differences between active and passive, straight and curved, become impossible to maintain. When a ship is veering, is that ship pushing against the waves, being pulled by them, deliberately steering, or accidentally? Consider for example the phenomenon of adaptation. We all think we know what that means. But on reflection, adaptation is a complex and curious event. An evolving species is adapting to another evolving species, since what we call rather glibly 'the environment' (that which veers around) is composed of nothing but other lifeforms and what one Darwinist calls their 'extended phenotypes', the results of their DNA mutations, and that of their symbionts, such as spiders' webs and beavers' dams.[8] A moving target is adapting to a moving target, which in turn is caught in a constantly morphing adaptation space. By definition, this process simply cannot be 'perfect', because *perfect* means that motion stops – but adaptation just is movement in adaptation space, and perfection would mean the end of adaptation, which is functionally impossible as long as evolution, which is to say lifeforms, continue. So when we talk about how lifeform *x* is 'perfectly adapted' to the swirl of phenotypes – including those that are 'its own' such as its also-constantly-evolving bacterial microbiome – we are saying something absurd, something on the level of *squarely circle*. We are trying to *contain* or *stop* the veering of attunements of lifeforms to one another, if only in thought. Teleology – the idea that things happen in line with some kind of end goal (or, by extension, that the ends justify the means) is the gasoline of 'perfect adaptation', and teleology, namely Aristotelian concepts of species development and depletion, is precisely what Darwinism liquidates.[9]

The phenomenon of adaptation should be sufficient to force us to recognize that attunement is the mode in which causality happens. Causality at all: a ball hitting another ball, a photon incident on a crystal lattice, an army invading a territory, the stock market plunging. As before, consider what happens when an opera singer's voice attunes to a wine glass. If done with the greatest accuracy, the wine glass explodes.[10] Think of how in the Paleolithic Era, painting or dancing a nonhuman was considered part of the process of hunting the nonhuman. The shaman follows the movements and habits of the prey, bringing them into her or his body, allowing his or her body to resonate with nonhuman capacities and qualities. Humans aren't necessarily Pac Man-like beings that munch everything into nonexistence – a fashionable way of thinking over the past few hundred years of modern Western philosophy (especially the dialectical philosophy of Hegel). Humans are sensitive chameleons.

We find a special and revealing adaptation mode in the syndrome we call *camouflage*. An octopus takes on the palette of the surface on which she is resting. A stick insect disappears into the foliage, to avoid predators. And at a basic level, to be alive is to adapt, without disappearing completely – to be protected by one's attunement, but not to the point of dissolving altogether. These brief glimpses of how what appears 'only' aesthetic to our eyes should be enough to suggest that attunement is not a case of having a blank, block-like substantial being whose superficial qualities are tuned while the substance remains the same – like what we think happens when we tune a violin: forgetting that the strings and the wood and the curvature of the violin form a unit such that tuning the strings by turning the pegs at the top of the instrument is *not* like arranging the apps on a smartphone, because the 'platform' is being altered by the tightening or loosening of the strings. The way attunement is deep rather than superficial is why the legend has it that Buddha taught meditation as a form of tuning: just as a sitar string should be neither too tight nor too loose, so one's mindful focus on the meditation object – a mantra, your breathing, whatever single object is the focus of your meditation – should be alert but relaxed. The conversation between 'alert' and 'relaxed' forms a dynamic system that simply can't remain still: hence the phenomenon that many beginning meditators experience, that their thoughts are rushing, because they are simply observing the intrinsic, rather than superficial, qualities of mind as such – mind thinks (in the largest sense of that word), mind 'minds', just as the ocean has waves. Movement is intrinsic. This fact becomes especially interesting when the meditation object is mind as such: when mind tunes to mind. What is experienced here is not absolutely nothing, but rather a strange beingness that cannot be pinned down to a presence I can point at.[11] There is a very deep ontological reason for this: appearing (waves) is intrinsic to being (ocean), yet different.

A lifeform is like that must-have eighteenth-century equivalent of the iPod and Bose speakers, the Aeolian harp. It's a string instrument that you place in an open windowsill. It resonates to the breezes that veer around the house. The haunting, harmony-rich, phasing sound this attunement system produces is strangely contemporary, as if Jane Austen characters were listening to a drone piece by Sonic Youth while they sipped their tea and played cards and wondered about the intentions of Mr Bingley in *Pride and Prejudice*. But sipping tea and playing cards are also attunement systems, exemplifying in this case the upper-class mode of consumer performance, in which establishing and maintaining a certain sense of 'comfort' is the basic tone to which the system is tuning: everyone must feel at ease, disturbance to the status quo must be minimized. All aristocratic attunement is about drones, sustained tones that waver as little as possible. If converted into sound, the space of polite interaction would indeed resemble a Sonic Youth drone piece.

Or consider 'bohemian' or Romantic (namely reflexive) performances in consumer space – the top-level performance we call consumerism, which now engulfs all other modes.[12] Reflexive consumer performance is just like meditation, insofar as one is tuning one's experience: one primarily consumes experiences, which are always of the other, as in the phenomena of window shopping or internet surfing. It is correct to claim that this attunement is a kind of 'spirituality', exemplified in the use of drugs or the nomadic wandering of the flâneur or the psychogeography of the radical French Situationist of the late 1960s.

How you appear and what you are intertwine deeply. In every single-celled organism there is a chemical representation, more or less accurate, of the realm in which it is floating. A perfect match – exactly the same chemicals – would equal death, which in a sense is a term for when a thing actually and wholly becomes its surroundings. Freud's *Beyond the Pleasure Principle* is a startling consideration of this fact.[13] Copying, mimicry, influencing and being influenced by, being tuned and tuning, things we do all the time in our environments, with other people, as we grow and learn to be adults and participate in activities – something causal is happening when these attunements happen, which is why we think 'primitive' (not-us) people imagine that photographs are stealing their souls. Representing and doing aren't so far apart. If I take a photo of you, haven't I in some sense snatched a part of you? In one way I have done so quite literally – photons that got influenced by your body, reflecting off it, have landed on my lens as I pressed the shutter.

Perhaps photographs really do steal your soul. Or perhaps photographs show you that your soul isn't yours in the first place, and that it certainly isn't inside you like a vapour in a bottle. The realm of attunement is thus like the mesmeric realm of animal magnetism. It is a force simultaneously discovered and repressed at the inception of modernity. When in the film *Dark City* the protagonist finds out that he can 'tune', what this means is that he can telekinese: he can do spooky action at a distance (Einstein).[14]

While modernity allowed agricultural logistics to destroy Earth even more successfully than it had done beforehand, it also unleashed, ironically and unwittingly, the non-agricultural ('Paleolithic') idea of an interconnective, causal–perceptual aesthetic force. Phenomenological and hermeneutic philosophy (some of the ingredients of this book) rediscovered attunement – more on this in a moment. Modern humans have recently rediscovered nonhuman beings outside the flattening, reifying concept Nature, which almost seems to have been designed to dampen our awareness of attunement space, perhaps just as the 'well-tempered' keyboard is designed to reduce the spectral harmonics that haunt a sound owing to its necessary physical embodiment: there is no sound as such, no pure tones, only the sound of a string, the sound of a sine wave generator. Objects thus have what is called timbre, and this is not an optional extra. Appearance is like that: appearance is better thought not in an eye-centred manner as candy decorating a cupcake, but rather as an object's timbre – its solitary quality, made up of a vast array of internal and surface qualities, that make it what it is, while also connecting it to where it is and the other objects around it.

We have rediscovered the veering brotherhood and sisterhood of nonhuman beings, once smoothed and packaged as Nature and indeed as 'the environment'. Kinship, as in sisterhood, as in *humankind*, precisely has to do with an uncanny intimacy, which is why the manufactured humanoid, the 'replicant' Roy shouts that word ('Kinship!') as with one arm he lifts his bleeding enemy up on to the roof of the tall building at the end of *Blade Runner*, an enemy who is his own kind (unbeknownst to both), in the form of agent Deckard.[15]

So in a weird ironic twist, humanity's flight from 'veering', which is the flight from our material embodiment, the timbre that haunts us with our affinities to chimpanzees, fish and leaves trembling on the tree outside the window, has ended up in a return to veering. Hegel describes the way history works via its shadow side, when he announces that the Owl of Minerva (the totem of justice, symbol of Athens) flies at dusk.[16] But Hegel's Owl didn't just fly at dusk. She flew straight out of a dream into the dreams of sleepers convinced they had woken up from every last trace of the so-called primitive. When we study attunement, we study something that has always been there: ecological intimacy, which is to say, intimacy between humans and nonhumans, violently repressed with violent results.

To begin to track this flight, then, is to veer towards a veering. And the first question that we might ask, in the context of an essay collection on veering and ecology, is whether nouns really are uninteresting until we make them more like verbs – give them the potential for action or force – because nouns denote things, and things are static entities that underlie appearances, which I have been arguing is fundamentally motion. Consider this noun: *future*. The future, or what Derrida called decisively *l'avenir* – the radically open future that is a possibility condition of the predictable future: is this term *future* unmoving? What happens at the end of this sentence? Does its meaning arrive? Arrive fully? This sentence means something, but you don't quite know what yet, as if the meaning, which is to say the tone to which it is tuning, were lying just off its end, elephant, seaweed, gamma ray burst. Is the future a thing? What is a thing? Haven't we already smuggled in a basic, default ontology before we start to think or talk about thinking, if we say *thing* is a noun and *noun* is static, and must be put in motion to be worthy of inclusion in a collection of essays on veering? All the objects in the world must be rounded up and forced to march and march until they drop, because that's the kind of work that makes them free?

Underlying this, isn't there a binary between *moving* and *staying still*, one that underlies most default (and incorrect) mechanical causality theories? A binary, moreover, that is part of the built, social space of Neolithic agriculture that eventually required carbon emissions to reproduce itself? A foundational drive (static or in motion?) to limit (a verb, therefore good?) ontological ambiguity (a noun, therefore suspect?), a drive that is structural (which side are adjectives on?) post-Neolithic social space, with its drastic tempering of attunement space into a sepia, anthropocentric consistency with the telos of 'survival'.

Object-oriented ontology is a way of thinking that wants to re-confuse us, much like deconstruction, about the status that we take for granted. Language as such is part of this taken-for-granted world, and myths of the origin of writing talk about how it is the bad neighbour, the uncanny weird sister of speech, that motile, fluid, 'living' force field that the legend says connected us face to face before the Fall, before the modern city state. Neolithic society, with its Linear A and Linear B (forms of writing) accounting for cattle – long lists of nouns, like receipts or bookkeeping – is an autoimmune disorder concerning its own ontological protocols. It reduces the world to anthropocentrically scaled manipulable stuff, and it doesn't like what it's doing – isn't that the content of agricultural-age religious origin stories? We have to farm now, and farming sucks, it separates us from the beasts and from our own life beyond survival (toil and sweat), but we do have to farm. We don't like the undead motility of writing, its spectral differentiation and deferral – the way it exceeds just accounting, just making lists of stuff that you own. What we like are clean boundaries between writing and speech, my field and yours, Heaven and Earth, God and Man, human and nonhuman (otherwise known as Nature), king and peasant, verb and noun. But the columns of double-entry bookkeeping tell us something about accounting, which never stops, unlike our fantasies about zeroed-out invoices and bottom lines. As credit attunes to debit, debit is attuning to credit: we have an intrinsically dynamic attunement system. Sentences never completely zero out. These phenomena are perhaps why we say that only two things are certain, death and taxes.

If writing really was invented during the agricultural age (and at least a certain very recognizable kind of writing was indeed invented then, the discourse of accounting), writing is a double-edged sword, a poison and a cure (pharmakon) as Derrida liked to argue.[17] Writing seems to be part of the drive to manipulate, to codify and assemble for easy demarcation purposes. But the very attempt to do so – and OOO would argue that this is because it is *ontologically* impossible, let alone linguistically impossible – summons the spectre of the radically free, unmanipulable playfulness of things, now observed slipping and sliding around within language as such, as the carefully ploughed intentions go haywire, as contradictory weeds begin to force themselves through the cracks in the social and philosophical sidewalk. *Pest*: is it a verb, or a noun? Do we need some kind of philosophical insecticide or herbicide to spray on words and objects to get rid of their ambiguous penumbra, their bacterial films, their trickster quality? *Trick*: is it a verb, or a noun?

Language doesn't want to stay put. Why? OOO argues that this is because things in general won't stay put, even when they are to all intents and purposes utterly still.

A perfume veers deliciously between verb and noun: the 'notes' emerge in time, so that a perfume evokes the future – how long will it stay? Into what will it be transformed? Perfumes tantalize because they veer and attune to human skin. Perhaps it would be better to think of OOO objects as *all* perfume-like in this respect. The word *essence* can mean intrinsic being and intrinsic flavour both at once, which is why perfumes can be called essences, as in *essential oil*. Perfumes attune to skin, releasing different smells at different moments depending on who is wearing them. We let perfumes veer, and they make us veer towards or away from them. Why not pencils, ash and star-nosed moles?

Let us then begin by veering between these reified categories. Reified, objectified, turned into a mere 'thing' – there we go again, we just think being a thing must be the worst fate imaginable. OOO's use of *object* is a mirror in which you see reflected your own prejudices about what objects are. If you think they are objectified, static, manipulable lumps of pure extension decorated with accidents – if you think they need verbs to get them going or adjectives to make them pretty – then you are going to think I am quite a mixed-up person.

Hesitation. Is it a verb or a noun? Is it movement, or stillness? Veering hesitates: in a way, hesitation is a quantum of veering. So we will start this veering process by seeming to veer off-course altogether, hesitating to begin. And you may find yourself examining the protocols of the discipline we now call sociology, and you may say to yourself, how did we get here? And will this rhetorical veer, this *assay of bias*, as the *Hamlet* character Polonius puts it, this curveball, catch the slippery *carp of truth*, as Polonius also puts it?[18] If it does catch something, would it be more like *seizing* (thrusting your hands into the cold water, grasping a fish) or *being seized*? In that latter sense, I'm mesmerized by the little fish, I track its darting movements with my eyes that begin to dart, fish-like, in their sockets. When I catch a fish perhaps I need to have been caught by it.

Acting Un-Civilized

Max Weber was one of the pioneers who inaugurated the discipline of sociology over a century ago, but sociology's structuring principle excludes the foundational concept on which it is based: charisma. Weber argued that societies based around charismatic authority – leaders that emit a sense that they have the ability to lead because of some inherent power (think of early Christianity or Islam) – give way to 'disenchanted', bureaucratic societies: modern European states, for example. But sociology does not explore enchantment. *Sociology itself is disenchanted*, and acts just like the bureaucratic society that Weber argues is disenchantment's birthplace; sociology is part of the logistics of what Weber called 'the disenchantment of the world'.

Sociology is afraid of its founder's concept, which was a little scary at the time too, since charisma has to do with forces that many described as supernatural or paranormal. Weber, like Kant, was himself fascinated by the paranormal. In general, one can think of modernity – world history since the later 1700s – as a profoundly awkward dance of including and excluding the paranormal. Freud, for instance, developed his theories as a way to bowdlerize the theory of hypnosis, which was in turn a bowdlerization of the idea of animal magnetism, Mesmer's hypothetical force (as we were seeing). Marx argues that capital makes tables *compute* value as if they were even weirder than the dancing tables of the quasi-religion of spiritualism that appeared to move when a spirit possessed the medium.[19] And so on – examples of this secret, almost completely untold history of modernity – of the tension-filled dance between the known and the unknown, the seeable and the unseen, the normal and the paranormal – are everywhere once you start to look.[20]

The paranormal is what religion was already excluding, religion being the way Neolithic society around 12,000 years ago monopolized what Weber calls charisma, taking it to mean a quality inherent to those that are already powerful, restricting it to the King who has the direct line to the god whom he hears ringing in his ears, telling him to tell the people what to do, *what to do* never being 'dismantle agricultural society, which has created patriarchy and tyranny in the name of sheer survival, and return to hunter-gathering and a less violent, less hierarchical coexistence with nonhuman beings'. Because that would be absurd. Heaven forbid we drop the anthropocentric equal temperament by which everything else becomes keyed to our teleological reference tone. That would be ridiculous primitivism – right?

Well, we are all still Mesopotamians. We are Neolithic humans confronting the catastrophe wrought by the Neolithic fantasy of smoothly functioning agricultural logistics, and we want to hold on to the philosophical underpinnings of those logistics for dear life, because otherwise . . . Well, it's unthinkable, it's woo-woo New Age obscurantist neo-fascist primitivist (find some more kitchen sinks to throw in here . . .).

This raging wall of resistance is directly proportional not to how impossible or difficult such a dismantling would be, but rather *to how easy it is*. It's not as if we would be giving up all control over our lives. We would just have different kinds of control. We could still do agriculture, for example – plenty of hunter-gatherers do; you don't have to cleave to the Neolithic model.

It is also easy because the logic underpinning Neolithic logistics is very obviously (when you study it) riddled with unsustainable paradoxes that result in cognitive and social violence (in the conventional sense, between humans) and ecological violence (in the conventional sense, regarding nonhumans). Equal temperament is riddled with awkwardly cramped and fudged frequencies, precisely to eliminate 'beating', the production of rhythmical pulses between tones, because the human manipulator of the instrument should be in charge of beating it according to what the human telos of the tune happens to be. It is biologically true that we aren't totally Neolithic – we have three-million-year-old bodies infused with Neanderthal DNA, and so on – but it is also philosophically and politically true. Because it is never true that there can be a perfect adaptation to one's phenotype, such that the search for perfection, now visible in seeds genetically engineered for tolerance to pesticides, must be destructive on numerous levels. Equal temperament dampens the haunting harmonics of an instrument's timbre, monoculture dampens biodiversity, logocentrism dampens the play of the signifier . . . and the dream of 'ecological' society as immense efficiency (the fantasy of perfect attunement) dampens the uneasy coexistence of lifeforms. We think we don't like veering – until an electric guitarist bends a note.

It is not just the upholders – the benefactors – of Neolithic society who are steering us away from veering. Those on the supposed other side of the fence – the so-called deep ecologists and the anarcho-primitivists – are only perpetuating agrilogistics and its devastating Nature concept, the idea that humans and nonhumans are profoundly different, based on needing to categorize human social space as a war against such things as 'weeds' and 'pests' – because their theories of a radical distance from the norm are still based on this norm *being* the norm. These approaches still work alongside a duality. Such concepts of difference as a rigid separation between humans and nonhumans are intrinsic to agrilogistics, the survival-at-any-cost strategy that began in the early Holocene and that has given rise to the feedback loops we now recognize only too well via the Sixth Mass Extinction Event, namely the fact that, among other things, 50 per cent of what biology calls animals (as opposed to fungi and viruses, for instance) have been wiped off the face of Earth in the last fifty years, because of anthropogenic global warming. Ecological thought requires *a different kind of difference*. Surely it's obvious that a slug is different from a panda. But it's different in the way a distant family member is different; not different in the sense of black versus white, or here versus there, or good versus bad.

It is *too easy* to dismantle the philosophical basis of our 'world' (aka 'civilization'). Without this basis, that world would collapse. The only thing inhibiting us is our habitual investment in that world, visible in the resistance to wind farms – we like our energy invisible, underground in pipes, so that we can enjoy the view. The very mention of changing our energy throughput raises the spectre of the constructedness of our so-called Nature. Think of the birds the turbines will kill! (Think of the entire species wiped out by *not* having the turbines.) (Are birds 'perfectly adapted' to oil pipes?) Think of the dreams we will be disturbing! We spent all this time tuning the world to anthropocentric tones, then delimiting attunement space. We might have to teach birds to tune to wind turbines, and this will be a drag. We want to be comfy in our unwavering, thanatological world.

Death is comfy, as Freud observed: the tension between a thing and the beings that veer around it is lowered to zero. A cell wall is ruptured and the cell's insides spill out into its surroundings. A glass shatters and the difference between itself and the space around it collapses. It's *life* that is disturbing and uncanny, all those energies flowing around, exchanges happening between the inside and the outside of an organism, exchanges between organisms, in every possible physical and metaphorical (and metaphysical) sense. Death means either totally not existing at all, or going around and around exactly the same all the time, like a perpetual motion machine – here's what I mean. Pop songs often have plenty of death in them, because death is very smooth and goes down easy; when I say *death* in pop, I mean the obvious, dull to some four-to-the-floor rhythms, the regular rhymes, the easy-to-hold-on-to earworms. As an artist you can either cheat death, or you can become death. Many a pop singer is death incarnate, because death always goes to number 1. Don't mistake those upbeat lyrics and dancey tunes for life. The frantic, maniacal repetition is exactly what Freud calls *death drive*. That kind of song is trying to parcel life energies into nice neat death packets. From this point of view, the trouble with some pop isn't that it's low or bad taste. It's that it's death warmed up enough (but not too much) to be a bit tasty. Death is powerful and compelling; life is fragile and shivery. Cancer cells are maniacal and can reproduce (repeat themselves) much, much better than normal cells – they are *more* alive than normal cells, which is why they kill you.

The funny thing is, if you try to avoid death, you can end up bringing it on. Think about eating burgers and fries literally to avoid the bitter taste of tannins. Bitter is a taste that infants have, without cultural training – they can all make the wincing face of tasting bitterness from birth. Bitter is a sign of poison. But in low doses, some poisons are essential – think of vitamins, too much of some of them can make you very seriously sick, but if you avoid them altogether, you also get sick. Perhaps you choose to eat burgers because you don't like that bitter taste. So you die more quickly of a heart attack or a stroke. Life is a balance between completely avoiding stuff and dosing yourself with stuff over and over again.

Many of our maniacal compulsive activities – such as washing our hands with soap all the time, and nowadays antibacterial soap – is precisely what brings on death in various ecological forms (such as upgraded superbugs). The maniacal flight from death *is death*. That's the weird feedback loop our kind of society is in.

Dismantling the underpinnings of agricultural logistics involves dismantling the 'metaphysics of presence', the idea that to exist is to be constantly present: to exist is to be a lump of extended stuff underlying appearances. Reality is a plastic, unformatted surface waiting for us (humans) to write what we want on it: 'Where Do You Want to Go Today?' (the

1990s Windows ad); 'Just Do It' (Nike); 'I'm the Decider' (George Bush); 'We create realities' (Iraq War press conference, 2005). There is the regular flavour of this metaphysics, basic default substance theories. We scholars all think we are superior to them, but they shape our physical lives which we happily reproduce, and we retweet them in the cooler flavoured upgrades, which speculative realism calls correlationism, which is the Kantian (and post-Kantian) idea that a thing isn't real until it has been formatted by the Subject, or History, or human economic relations, or Will to Power, or Dasein . . . In a way it's a worse (in the sense of more ecologically destructive) version of the regular substance ontology flavour. This is because it treats things not like lumps, but like blank sheets or screens. Lumps are at least three-dimensional. Imagine arguing, as some do, that there are only blue whales when we *say* there are (they are cultural constructs, they are discursive products of epistemic formations, they are concepts we project onto certain lumps of marine matter) . . . And lo, it came to pass, there were no longer any blue whales . . .

OK, so happily that particular extinction hasn't yet occurred. It hasn't yet occurred because people became enchanted by recordings of whale sounds in the mid-1970s. Enchanted. What does it mean? In terms of charisma, it means some of us submitted to an energy field emitted by the sounds of the whales. The fact that in my line of work (the academy) this is a wholly unacceptable, beyond the pale way of describing what happened is a painful and delicious irony. You can't say things happen because of vibes. That's what hippies say. And we're not hippies. We're black-clothes-wearing cool kids who wouldn't be seen dead in what one comedian calls 'multicoloured, ill-fitting, vaguely ethnic clothing'. We spend most of our time trying not to sound like Yoda.

Just as attunement is the fuel of veering, so charisma is the fuel of attunement. Charisma makes us hesitate, wavering in its force field.

What if charisma were *actual*? What would the emission of such an energy field imply? It would imply, for a start, that art isn't just decorative candy. It would imply what 'civilized' philosophy from Plato on has been afraid of, the fact that (shock horror) art has an effect on me over which I am not in control. Art is *demonic*: it emanates from some unseen (or even unseeable) beyond in the sense that I am not in charge of it and can't quite perceive it directly, in front of me, constantly present. A dangerous causative flickering: magic. Magic is taboo cause and effect, or unthinkable cause and effect: either ridiculous or dangerous or impossible, or some weird borrowed-kettle combination of all three. (How can something be impossible *and* dangerous?) What we are talking about is what Einstein called spooky action at a distance, by which he meant quantum entanglement, but which also means what happens if you try to visualize the Rothko Chapel even if you aren't there, even if you have never seen the Rothko Chapel in real life, or perhaps even if you have never actually seen a Rothko painting, or a postcard of a Rothko painting, but have only heard about Rothko.

The Rothko Chapel, a non-denominational space in central Houston, is one of Mark Rothko's final works, and it's located just behind where I live. It's a cool, dark space where the walls are adorned with gigantic versions of Rothko's characteristic abstract fields of vibrating colour, in a range of dark purples, blues and blacks. We might conventionally argue that the charisma of the Rothko painting is bestowed upon it by humans: this would be the acceptable Hegelian way of putting it. We make the King be the King by investing in him. Investing what? Psychic energy – which, if you recall, is a bowdlerization of the Force-like animal magnetism. What if this attitude were not only masochistic in the extreme, but also – incorrect? After all, as Schrödinger has already argued, the one thing you can rely on is that at the very least two tiny things (an electron, a photon) can be 'entangled' such that you can do something to one of them (polarizing it, changing its spin), and the other will, for instance, polarize in a complementary way instantly – which is to say, faster than light. And this complementary behaviour happens at arbitrary distances. You can now observe two particles separated by kilometres behaving this way; one is on the other side of town; one is on board a satellite, and so on – arbitrary means 'even if that particle is in another galaxy'. And now physicists are experimenting on scales trillions of times bigger than electrons and photons, with positive results. For example, you can entangle clusters of carbon atoms called 'buckyballs' because of their shape, which resembles Buckminster Fuller's geodesic domes. You can smear a tiny but visible tuning fork into quantum coherence, where you can see it (with the naked eye) vibrating and not-vibrating at the very same time. You can cool a tiny but visible mirror down towards absolute zero and isolate it in a vacuum – in other words, nothing at all is pushing it mechanically – and you can observe it shimmering, moving back and forth without mechanical input.[21] And there are to date no loopholes; there is not some underlying substance that means the two particles are really one, for instance.[22] The entities are different. But they're not totally separated. (This idea resembles what I was saying a little bit earlier about ecological thought requiring a different kind of difference.)

Causality just is magic. But magic is precisely what we have been trying desperately to delete.

Magic implies causality *and* illusion, and the intertwining of causality and illusion, otherwise known in Norse-derived languages as *weirdness*. *Weird* means strange of appearance, and it also means having to do with fate.[23] Neolithic ontology wants reality not to be weird. Eventually weirdness is confined to Tarot cards and vague remarks about synchronicity. What does it mean, though, to entangle illusion and causality? What it means is that how a thing appears isn't just an accidental decorative candy on an extension lump. Appearance as such is where causation lives. Appearance is welded inextricably to what things are, to their essence – even 'welded' is wrong. Appearance and essence are like two different 'sides' of a Möbius strip, which are also the 'same' side. A twisted loop is exactly what *weird* refers to, etymologically speaking. The minimal topology of a thing is the Möbius strip, a surface that veers all over, where a twist is everywhere. This is because the appearance of a thing is different from what it is – yet the appearance is inextricable from it, too. There is no obvious dotted line between what a thing is, and thing-data. Attuning is like studying a Möbius strip, a special object in the form of a twisted loop. It's not hard to make one: tear a thin strip of paper, twist it and join the ends together. You will see that when you trace your finger around the shape, you land on the 'other side' to where you started, but you never actually 'flip' to an 'other' side. This is weird; it means that the shape only has one side.

Unfortunately for the scientistic ideology that dominates our world and the neoliberalism that forces us to behave in scientistic ways to ourselves, one another and other life-forms, *the idea that appearance is where causality lives is also just straightforward modern science.* Hume's argument was precisely that when you examine things, what you can't see directly is cause and effect. All you have are data, and cause and effect are correlations of those data. So that you can't say 'humans cause global warming' or 'cigarettes cause cancer' or 'this bullet you are firing at point blank range at my temple will kill me'. You can say 'It is 97 per cent likely that . . .' – thus opening the door to the deniers, who are in fact modernity deniers, unwilling to let go of the clunky mechanical, visible, constantly present causality that you can point to.

And as we have seen, Kant underwrote this devastating insight: all we have are data not because there is nothing, but because there *are things*, but these things are withdrawn from how we grasp them. Kant's example: raindrops fall on your head, they are wet, cold, spherical. This is raindrop data, not the actual raindrops. But they are raindrops, not gum-drops. And they are raindroppy: their appearance is entangled with exactly what they are.[24] What art gives us, argues Kant, is the *feel* of data, the data-ness of data, otherwise known as givenness. This data-feel is, he argues, an attunement space, the one place in the whole universe where mesmerizing hesitation can happen – a very important mesmerizing hesitation, because it underwrites the existence of a priori synthetic judgement, because in this experience I get a magical taste of something beyond my (graspable) experience, a transcendental beyond-ness that Kant wants to restrict to the transcendental subject's capacity to mathematize. But Kant's analogy – he was afraid of analogies for just this reason – is the raindrop, and the raindrop's mathematizable properties such as size and velocity are also, he states, on the side of appearance.

The aesthetic dimension is a necessary danger, a tiny bowdlerized zone of mesmeric attunement, without which we couldn't know that there is a weird gap between what things are and how they appear, which is why we know we should treat the beings we call *people* as ends and not as means, because your uses of me never exhaust me in principle. Through this attunement, I get to discover that my inner space is infinite, like Doctor Who's TARDIS. But it's equally likely, according to the implicit logic of Kant's *Critique of Judgement*, if not its explicit argument, that the beautiful thing is *also* bigger on the inside. The experience tells us that maybe *everything is a TARDIS*. For the beauty experience is precisely that phenomenon in which I find it impossible to tell who started it: was it me or was it the thing? Yet Kant concludes that this secretly means, *we (the subject) started it.* That means that I'm not especially different from other things, such as crickets and even cricket bats. But it also means that crickets and cricket bats are kind of special in the way I think I myself am, as a person. It means therefore that cricket bats might be a little bit 'alive' in some way, and crickets (and maybe the bats too) might be 'people' in some way. Crazy, right? No wonder you're not allowed to say this out loud in a university cafeteria.

A small piece of mesmeric, magical dynamite is embedded at the crucial point in modernity's architecture: a tone that is rich with harmonics 'disgustingly' not quite keyed to human teleological reference frames. The self-transcending subject is underwritten by a mysterious power emanating from the non-subject (the 'object'). I may be the one who gets to decide whether the light is on in the refrigerator or not (correlationism); but there needs to be a refrigerator in the first place, and for some reason I find myself drawn to it. The Sami people of northern Scandinavia have long been oppressed by corporate greed and nation-building. Why then are the Sami people reluctant to cast counter-spells to those woven by global corporations? Because that would involve their culture with corporate culture in a mutual attunement space: their culture would be distorted by the attempt.[25]

People are Strange When You're a Stranger

Things are exactly what they are, yet never as they seem, and this means that they are virtually indistinguishable from the beings we call *people*. A person is a being that veers in just this way. Once we start embracing difference not as rigid separation but as uncanny affinity, as I've been suggesting, we see that humans are more like nonhumans, and nonhumans are more like humans, than we like to think – and those two phrases do not quite add up. It is radically undecidable whether we are reducible to nonsentient, nonconscious, non-person status – or whether things that aren't us, such as foxes or teacups, are reducible upwards to conventional person-hood. I might be an android – this android might be a person: that's the best we can do. Deleting the hesitation by reducing either one to the other is what is called violence. If I decide you're just a machine, I can manipulate you exactly as I want. If I decide you're a person, and person means 'not a machine', then I can decide that other things are just machines by contrast, and manipulate them.

I am playing a tune called *myself* to which you are attuning, but which is itself attuned to you, so that we have an asymmetrical chiasmus between *myself* and me, between me and you.[26]

We live in a world of tricksters. How we conduct ourselves in this world, the ethics of the trickster world, has to do with respecting that subjunctive, hesitant, might-be quality. It has to do with attunement. As I was saying before, in the context of thinking about life, attunement is a dance between completely becoming a thing, the absolute camouflage of pure dissolution (one kind of death) and perpetually warding off that thing (another kind of death), the mechanical repetition that establishes walls, such as cell walls. Between *I am that* and *me me me*: in other words, between being reducible to other stuff (I'm just a pile of atoms or mechanical components) and being totally different from other stuff (I'm a person, and only some beings get to be people). What is called life is more like an undead quivering between two types of death, a deviance that is intrinsic to how a thing maintains itself, or metastable as some like to say. Some things need to deviate to stay the same. Think of how a circle is how a line deviates from itself at every point, thanks to the seductive force of a number existing in a dimension perpendicular to that of the rational numbers (pi).

My experience of showing guests around Rothko Chapel has provided me with beautiful examples of how, if you're scared or critical of art (perhaps you have been taught it's always a product of political oppression, or bourgeois sensibility, or a mystification designed to confuse you, or something like that), you find the sort of attunement that happens in there very uncomfortable. It's because you can't shrug it off or dismiss it as some unreal, ideological effect. Something is really happening – oh no, get me out of here! Because the Chapel is 'religious', you can't just put the paintings in a box with the label 'art'. Because the 'religious' quality is not specific, but more like a free-floating 'spirituality', you can't put that in a conceptual box either. Religion is turned into something like appreciating art; appreciating art is turned into something like spiritual contemplation. And those two transformations don't neatly map on to one another. So you can't dismiss what you're feeling as purely a social construct quite so easily.

The upshot? Some scholars have only lasted two minutes in the Rothko Chapel. Some other friends, such as Björk and Arca, another musician friend, stay in there for ages and ages, soaking it up.

Why is this feeling of attunement scary for some? It's because it appears not just to be something they're in charge of, but something that's emanating from the paintings and the space itself. We attune to the gate-like rectangles of aubergine space, because they are already tuning to us, waiting, beckoning. A Rothko Chapel painting is a portal: just what might come through? Such a painting is a doorway for what Derrida calls *l'arrivant* (verb or noun?), the *future future*, the irreducible, unpredictable one. Philosophy, which is wonderment (hence horror, or eroticism, or anger, or laughter) in conceptual form, is an attunement to the way a thing is a portal for the future future. The love of wisdom implies that wisdom isn't fully here, at least not yet. Perhaps if it ever succeeded in teleporting down perfectly, it would cease to be philosophy. Thank heavens philosophy *isn't* wisdom. If it is, I want nothing to do with philosophy.

We might want to contain the aesthetic experience by framing it as 'art' in some predictable, preformatted sense. Going further, we might think art is a reflex of the commodity form, which would really help us to keep our suspicious distance: heaven forbid we be seduced by anything. Art shows us how a disturbingly ambiguous pretence is woven into aesthetic experience: wonderment is based on the capacity to be deceived. The more we are OK with being lied to, the wiser we might become. 'Ever get the feeling you've been cheated?' (John Lydon, aka Johnny Rotten, once said onstage during a Sex Pistols concert). So perhaps we could dismiss a Rothko painting, as art critic Brian O'Doherty does in his famous essay on the commodification of art space, the dreaded 'white cube' of the contemporary gallery, now replicated in a million minimalist townhome interiors.[27]

We want art to make us sure that we aren't being conned

or ripped off or pitched to or prostituted or sold to: tuned. But this is exactly what art can't do. Art theory in modernity tends to want to distinguish art from conning or selling or ripping off, and from the dreaded status of 'object'; and this results in art's confinement to a tiny experiential region, sophisticated beyond sophistication, purer than the white cube purity of the philistine buyer and owner, hanging on the white walls but above anything that smacks of gross consumerism.

As anyone who is vaguely familiar with the very high-stakes and high-priced art industry will attest, this abstinence (and abstinence from abstinence) is exactly the top level of consumer space: the self-reflexive, 'Romantic' mode of bohemian consumerism, in which we are all caught. Think of how we all like to say we no longer follow *fashion*, but instead select our very own *style*. One style can then be sampling everyone else's style, and this can seem as if you are floating above everyone else, the poor fools, trapped in consumerism. Yet this performance, which we could call 'I Am Not a Consumer', is the *ultimate* consumerist performance. O'Doherty has no time for what he thinks is the abstracting, reifying 'Eye' induced by the white cube space itself. But he has even less time for the poor corporeal 'Spectator', the comical, humiliated body dragged around by this eye.[28] O'Doherty is saying that the way art galleries are set up, we are moved around them in a passive way, watching ourselves from an abstract distance. This means that being passive is bad, because being passive means being an object, which means not being a subject. Heaven forbid that we become an object, heaven forbid that we ever become passive – that would be a fate worse than death.

Attunement is the feeling of an object's power over me – I am being dragged by its tractor beam into its orbit. And yet we are told that we are not to be manipulated. We write essays such as *Inside the White Cube* about how white cube spaces inevitably seduce us all – except for me, the narrator of the white cube essay, and you, the sophisticated reader whom the essay is interpellating, rising above it all, exiting the poor beastly body and the abject world of objects, like the Neoplatonic soul transcending the body. 'Obey your thirst' (advertisement for the soft drink Sprite, 1990s) has no effect on us. Everyone gets conned by objectification, except for me, the one who writes the sentence *Everyone gets conned by objectification*. All sentences are ideological, except for the sentence *All sentences are ideological*. Can you see how this works?

Critique mode is the mode of the pleasure of no-pleasure, the sadistic purity of washing your hands of the crime of being seduced, as if *detuning* were about exiting attunement space rather than what really happens, which is only *retuning*. In this mode, the worst thing that could happen is that you could make or enjoy kitsch. Happily, children have never heard of such things. My son Simon tells me that if you cross your eyes and stare at a Rothko painting just so, the red lines will start to vibrate and float towards you, and you will feel nauseous and giddy – and that these are exciting, oddly pleasant sensations, like spinning in a swivel chair. Apparently the paintings aren't just commodities sitting primly in a shop window. Apparently they even exceed their human-keyed 'use value'. For O'Doherty, the best kind of art, which he calls postmodern, is an endless conversation between (human) subjects about what good art might be, as if tuning up were not part of the orchestral performance – a myth rapidly dispelled by the first few seconds of 'Sgt. Pepper's Lonely Hearts Club Band'.[29] Actually letting yourself enjoy a thing is pleasurably avoided. And yet to a six-year-old child, it's obvious that Rothko is trying to blow your mind.

Art sprays out charismatic causality despite us. And unlike a lot of things in our current world, and within limited parameters (sophistication, taste, cost), we still let it in. Art is a realm of passion for no reason: I just like this particular shade of blue, I want you to feel the weight of this metal toe, come in to this installation, look, peer through the curtain. The time of novels is the time of lust – the first novels were necessarily pornography (Aretino). So when we talk about art, we are talking in the region of love and desire, those unsteady, uneasy, wavering partners.

Let us widen our gaze from the artwork to a more general description of this region. Love is not straight, because reality is not straight. Everywhere, there are curves and bends, things veer.

Per-ver-sion. En-vir-onment. These terms come from the verb *to veer*. To veer, to swerve towards: am I choosing to do it? Or am I being pulled? Free will is overrated. I do not make decisions outside the universe and then plunge in, like an Olympic diver. I am already in. I am like a mermaid, constantly pulled and pulling, pushed and pushing, flicked and flicking, turned and opened, moving with the current, pushing away with the force I can muster. An environment is not a neutral empty box, but an ocean filled with currents and surges. It environs. It veers around, making me giddy. An aesthetic wormhole, bending the terrestrial and ecological into the cosmological. The torsion of deep space, beaming into the cold water of this stream like bent light, the stream where I was caught by the fish I was catching a few pages ago.

Spacetime as such is a bending, a curvature. It isn't correct to say that spacetime is first flat, then distorted by objects. Objects directly *are* the distortion of spacetime: spacetime is the distortive force field that emanates from them. Curvature, lumps and bumps, a strange plenitude everywhere, no dead air. Spacetime isn't a flat blank sheet that gets disturbed. Spacetime is disturbance. A disturbing lens of matter-energy, we see as much as we can see, always less than all, through the convex kaleidoscope of spacetime. A thing is dappled with time. But not a lump coated with time, improved by the makeup of motion. Better: a thing *is* this temporal dappling.

The nineteenth-century writer John Ruskin was a great scholar of architecture who argued that the modern tendency to want to clean old buildings, very much in effect today, was a sacrilegious erasure of what he liked to call *the stain of time*.[30] In a sense Ruskin was aiming at something like an ontological redescription of things: to remove the time stain is to harm the actual thing, because a thing *actually is* this temporal staining. To want to cleanse a building of what is taken to be a supplementary stain is to assume that a thing underlies its appearance, the old default substance ontology. To allow things to get dirty is to allow that things are not at war with time. Further still, the 'dirty' Sistine Chapel ceiling painted by Michelangelo is similar today to how it would have been seen in flickering candlelight.

Newton's world is a realm of straight love, instant beams of gravity that are God's love, everywhere, all at once, outside time, the omnipresent force of an omniscient being acting on static extensional lumps, exciting them, pushing and pulling them around like cattle.

We do not live in Newton's world.

Einstein's world is a realm of perverse desire, invisible ripples of gravity waves that make up spacetime, the invisible ocean in which the stars float submerged. We love the dead. We love fantasies. Do they love us back? We are pulled towards them and as this happens, time expands and shrinks like a polymer. No God could be omniscient in such a world, where time is an irreducible property of things, part of the liquid that jets out of a thing, undulating. There are parts of the universe that an observer will never be able to check. They are real. Things happen there. But some observers will never know *where* they are happening, or *when* they are happening. Some people in the universe will never know you are reading this, because they never *can* know. Just as you won't be able to know them.[31]

In a universe governed by the speed of light, parts are hidden, withdrawn, obscure. The dark Dantean forest of the Universe, an underwater forest of rippling weeds. You should find this idea extremely comforting. It means that you cannot be omnipresent or omniscient. It means that you cannot look down on the poor suffering beings of the universe from a position outside time, and smile sadistically at their pain, a smile we often call pity. This is what we sometimes call the abstract gaze of the Enlightenment, that period in the early history of modern Europe and America in which universal values were articulated, unfortunately at the expense of urgent particularities such as race, class and gender. Many artworks of this period, such as C. F. Volney's *The Ruins of Empires* or Shelley's *Queen Mab*, are staged from precisely this position outside of the universe as a way to judge it.

Each entity in Einstein's universe is like the veering turbulence in a stream, a *world tube* or vortex that cannot know all. There is a darkness that cannot be dispelled.

Consider now the even stranger, and even more accurate, description of things we call quantum theory. In quantum theory, the binary between moving and staying still – between a certain concept of *verb* and *noun*, or between a certain concept of *object* and *quality* – becomes impossible to sustain. Objects isolated as much as possible from other objects still vibrate without being pushed, that is to say, without being subject to mechanical causation.[32]

The idea that I'm outside the world, looking in, wondering which choice to make, is the ethical equivalent of the substance ontology that separates being from appearance with firewalls and fungicides. But the traditionalist 'conservative' versions of this line of thought, called 'environmentalism', also try to contain wavering, the hesitation filled with the vibrations of attunement. It's called environmentalism, but it's not en-vir-onmental enough.

And this isn't surprising, because 'traditional' agrilogistics ends up as our current version, so that there is a line from the notion of the guiding weight of tradition to the play of infinite (human) freedom and 'choice'. The aesthetic dimension is commonly imagined as a special glue that sticks these two poles together, by allowing humans to impose the proper form, to adapt their world perfectly to their requirements. But this is not how it works. We have seen that this dimension is deeply entwined with things as such, not with (human) formatting. There is a certain courage of letting yourself fall asleep and allowing dreams to come, which resembles the courage of allowing art to affect you. Hallucinatory phantasms are a condition of possibility for seeing anything at all. Hearing is a chiasmic crisscross between sounds emitted by my ear and pressure waves perturbing the ear's liquids from the outside. The not-me beckons, making me hesitate.

Escape from the Uncanny Valley

When we examine lifeforms, we find that they are much stranger than we take them to be, in part because the concept of 'life' is much stranger than we take it to be. Biology, a discipline whose name was coined in 1800, is the science of life, and one of the results of being scientific about life is that it becomes more and more difficult to draw, with a straight face, a boundary between living and nonliving things. We are, after all, made of chemicals.

We've been talking about causal connectivity between lifeforms, how things are interrelated in what we sometimes call the web of life. Now let's take a look at something more in the region of aesthetics and ethics: how does connection *feel* and what does connecting *look like*?

Saying that things are connected doesn't necessarily mean the same as saying that they are totally mashed together. Things are *dependent* on one another: that means that some things might be more dependent on some things and less dependent on other things. It's a loose, wobbly system of connection, rather like a large model made of Lego or an unstable mobile where thin wires connect cut pieces of card that float above your bed. If everything was totally mashed, then connection would be no problem, either causally or ethically or aesthetically. But connection is a big problem. For instance, environmentalist charities are known for encouraging us to donate money by depicting what are sometimes called 'charismatic megafauna': big cute animals such as pandas. What about slime moulds or worms, or for that matter, bacteria? Global warming is tough for bacteria too in ways that might be disastrous for the soil and so disastrous for humans.

Whether we're thinking about how lifeforms look or how we behave towards them or what we know about how we relate to them in the realm of cause and effect, we are dealing with wonky, fragile systems. We seem to have developed a particularly lopsided way of perceiving lifeforms – the aesthetic part. It's not that our connection to them in this domain is like a flat plain with a smooth slope: at the top of the slope are lifeforms with whom we can identify, at the bottom are those who don't turn us on in the slightest. It's not a smooth slope. It's more like what in robotics design is called *the Uncanny Valley*.

What is the Uncanny Valley? Imagine you are standing on top of a hill. You are looking across a valley towards the hill on the opposite side. On the opposite side of the hill is a cute little robot, possibly the classic *Star Wars* character R2D2, or the more recent *Star Wars* cute robot BB8. The theory of the

Uncanny Valley says that these sorts of robots are cute because they don't remind you of yourself at all, so that when they communicate, you experience it as charmingly non-threatening to your sense of who you are. You don't feel like they're trespassing on your turf – the turf that says you're human and they're not.

Further away, behind the peak, are robots that have nothing to do with anything obviously human, such as industrial robots. You couldn't care less about them. They don't try to talk to you, but R2D2 is trying to talk to you. He is more along the lines of a stuffed animal toy. Then, as you look down into the valley, you see all kinds of beings that become more and more disturbing the longer you look. There are robots that are unnerving for their human qualities, their similarities to us. And there are corpses somewhere down there. And right at the bottom of the valley we have animated corpses, zombies. They are dead and disgusting and also disgustingly alive. Really lifelike puppets inhabit the slopes of the valley nearest to you, further up from the zombie level.

Somewhere in the Uncanny Valley are all the humanoid, hominid, hominin-type beings, beings we define as genetically close to us, or designed to resemble us. The theory runs that we are disturbed by them because they resemble us too closely. It has only recently been admitted, for example – because you can't really ignore DNA evidence – that not only were we, *Homo sapiens*, much closer genetically to Neanderthals than we like to think, but we had sex with them to the point where a significant sliver of our DNA right now is derived from Neanderthals. Again and again we have told ourselves the story that although Neanderthals are like us, they are enough unlike us at the same time for there to be a comfortable distance between us and their 'primitive' nature – but maybe the truth is we know that *Homo sapiens* are closer to Neanderthals than we would like to think, that we might even be slightly Neanderthal . . . and this freaks us out. We have also been telling ourselves that they are not as *sapiens* (wise) as we like to call ourselves. So that we 'had' to wipe them out, because they were basically getting in the way of our projects, which were far more forward-thinking than they could possibly handle. Or we tell ourselves that they couldn't have been conscious like us, because if they lacked a powerful sense of the future, they also had little or no imagination. That's why they weren't looking when we ambushed and exterminated them. This sounds a bit circular. We can prove that Neanderthals weren't that great, because we got rid of them, because they weren't that great. The irrational circularity has to do with how we think and feel about lifeforms, including ourselves, right now: the unconscious or semi-conscious or otherwise structural attitudes that shape how we behave in this moment.

How steep your valley is might be a good indicator of syndromes such as racism and speciesism. A very steep valley would indicate that you have done a lot of work banishing the uncanny beings to some nether domain of your thinking or feeling or awareness (or what have you). You are so freaked out by them, so disgusted, that you have almost forgotten them. Or you might be a bit more tolerant of them, and your valley might be quite shallow.

Yet whether it is steep or not, there is still a valley. You are still distinguishing yourself in some way from the beings in that valley. Why are they in there? I think the defining characteristic is *ambiguity*. Are they related to me or not? When I look at them, they seem to have recognizable features. But something about them is very strange: perhaps they are androids, for example, and if they are androids and I'm so like them, maybe I myself am an android. And this is what really disturbs me about them: I might have more in common with them than I think I want to. When you start to think this way, the valley becomes an artefact of anthropocentrism, racism and speciesism – of xenophobia, a fear of the 'other', which is, often, really a fear of what we have in common with the 'other'. A sneaking sensation that we are not as distinct from these robots and zombies – or people from different cultures or genders – as we like to say we are. This proximity is what causes that uneasy, uncanny feeling. Instead of recognizing it for what it is, most often we push it away, trying to keep the distance between our peak of distinction and the valley below.

We want to have clean, rigid distinctions between beings: it's called, quite rightly, *discrimination*. But just because something is distinct and different, doesn't mean we can distinguish it from us in some ethical or ontological way. That's the trouble with our poor Neanderthal. She looks like us, an awful lot, but she is pretty distinct too. She falls between categories.

If you go to your local natural history museum these days, you may see, as I do when I go to the one in Houston, a wall-sized graphic of many tens of lifeforms that are linked to us in the immediate history of the evolution of humans, from lemurs to ourselves. The graphic will be a sprawling, uneven net. It's like your family and this is like looking at a family tree. The lifeforms are like you, enough like you and enough connected to you for you to feel cognitively comfortable or at least familiar, as they say, if not comfortable all the way, like you can tolerate being at their house for dinner.

But precisely because of this, the lifeforms are also unlike you. Uncle John always had this disconcerting habit that really disgusts you. You have no idea why *she* is your sister, you might as well be from different planets. These beings are familiar and strange at the same time. In fact, the more you know about them, the more strange they become. You don't get rid of the strangeness by knowing more about a thing, necessarily. Isn't science a way of realizing that? Our universe is so much more strange now that we know more about it.

The word for *familiar and strange at the same time* is *uncanny*. We are racists and homophobes and sexists when it comes to beings who we put in the Uncanny Valley – because we put humans down there too. It's not exactly *otherness* that we are working with here. Ethics and politics might not be about tolerating, appreciating or accepting otherness. Ethics and politics might be about tolerating, appreciating or accepting *strangeness*, which boils down to *ambiguity*: how things can appear to be oscillating between familiar and strange, for example.

Doesn't appreciating art have to do with allowing things to be ambiguous? It's not just that there are all kinds of paintings and sculptures and books and pieces of music in this world, with all kinds of cultures to do with how these things are made, received and interpreted (and so on). What it is, and this is the most basic thing perhaps, is that you have no idea what this artwork will 'say' to you next: it's especially obvious when you've lived with a favourite piece for several years.

Deeper still, there is something strange that happens in the appreciation of art, which many philosophers have found disturbing. It's disturbing how the experience of relating to art, for example, makes it difficult – sometimes impossible – to sustain the valley across which we see other entities as 'other'. Let's see how. It's pretty obvious that art has an effect on me, and this effect is to a large extent unbidden: I didn't ask for it, which is part of the fun. I had no idea I could be affected in precisely *this* way. My whole sense of what 'affect' means has been transformed by this artwork – and so on. When I love an artwork, it is as if I am in some strange kind of mind meld with it, something like telepathy, even though I 'know very well' (or do I?) that this thing I'm appreciating isn't conscious, isn't sentient, isn't even alive. I am experiencing unknown effects on me coming from something that I am caught up with in such a way that I can't tell who 'started it' – am I just imposing my concepts of beauty on to any old thing, or is this thing totally overpowering me?

The real feeling of experiencing what we sometimes call beauty is neither about our putting a label on to things, nor of our being absolutely inert. Instead it's like finding something in me that isn't me: there is a feeling in my inner space that I didn't cook up myself, and it seems to be sent to me from this 'object' over there on the gallery wall, but when I try to find out exactly where this feeling is and what it is about the thing, or about me, that is the reason why I'm having this feeling, I can't isolate it without ruining what precisely is beautiful about it.

What is the difference between *tolerate* and *appreciate*? It is all about this theme of coexisting. *Tolerate* means that within my conceptual reference frame, I allow something to exist, even though my frame doesn't really allow it. *Appreciate* means that I just admire it, no matter what my reference frame is. That's why we use the term *appreciate* to talk about art. No one says 'I really tolerated that Beethoven string quartet' in a positive way. But you can easily say 'I really appreciated that disco tune' and people will know that you mean something positive.

When you think about it like this, you can see why being able to *appreciate* ambiguity is at the basis of being ecological.

And do you know what this means? Your indifference to ecological things is exactly the sort of place where you will find the right kind of ecological feeling. This is one big reason why deleting the indifference too aggressively and too fast, by being preachy, doesn't help at all. *You don't know why you should care*: isn't that what we are all feeling when we experience something beautiful? How come this chord sequence is making tears run down my face?

Reasons for being nice to other lifeforms abound, but around them there is a ghostly penumbra of feelings of appreciating them for no reason at all. Just loving something never has a great reason attached to it. If you can list all the reasons why you 'should' love this particular person, you are probably not in love. If you have no idea, you might be nearer the mark. This ambiguous spectral aesthetic halo around ethical decisions doesn't tell us how to act, or even whether or not to act. It has a 'passive' quality about it, as if even our distinction between *active* and *passive* were not that thin and rigid, and that what is often meant by *passive* is in fact the penumbra we are talking about. Is how you relate to a beautiful artwork active or passive? You certainly don't want to eat it, because that would get rid of it, and you like it. But it's not clobbering you either. It's affecting you, but in a non-violent way.

When you tolerate another lifeform, it's like leaving them in the Uncanny Valley, although you admit that you need to go down and help them – returning afterwards to your peak. When you appreciate a lifeform, for no good reason, it's as if you made the Uncanny Valley a bit shallower. If you carry on like that, the Uncanny Valley starts to flatten out. It flattens out into something I like to call the *Spectral Plain*.

What is the *Spectral Plain*? It's a region that seems totally flat, and it extends in all directions. And on this plane, I can't distinguish very easily between *alive* and *not alive*, between *sentient* and *non-sentient*, between *conscious* and *non-conscious*. All my categories, which excavated the valley, start to malfunction. And they malfunction deeply. If they just went away, I would have my answer: I would be able to *collapse* life into nonlife, for example, so that really, there are no living beings, just bunches of chemicals (this is a popular materialist reductionist solution to the problems of knowing a lot). According to this, the malfunction can be fixed: I can eliminate ambiguity. In that case, what was wrong with the Uncanny Valley was precisely that it made me feel ambiguous.

But I don't think that's what's wrong with the Uncanny Valley. I think what's wrong with the Uncanny Valley is *the peaks on either side of it*. What's wrong is that we aren't in it ourselves – nor are the robots we like to think of as lovable toys. Remember, this is an experiential valley where beings such as zombies live in between peaks: we 'healthy' humans live on one peak, and all the cuter robots on the other. Zombies live in the Uncanny Valley because although they embody Cartesian dualism of mind and body, which is how we like to think about ourselves, they do so not in the standard, 'nice' way: they are animated corpses. It is as if they are mocking this dualism – they are a parody of this dualism – as if when we look at them we have a fantasy concept that shows us that there is something actually very wrong with mind–body dualism.

The Uncanny Valley concept explains racism and *is itself racist*. Its decisive separation of the 'healthy human being' and the cute R2D2 type robot (not to mention Hitler's dog Blondi, of whom he was very fond) opens up a forbidden zone filled with uncanny beings that reside scandalously in the Excluded Middle region. The distance between R2D2 and the healthy human seems to map quite readily on to how we feel and live the scientistic separation of subject and object, and this dualism always implies repressed abjection as we have just seen. R2D2 and Blondi are cute because they are decisively different and *less powerful*. It is this hard separation of things into subjects and objects that gives rise to the uncanny, forbidden Excluded Middle zone of entities who approximate 'me' – the source of anti-Semitism to be sure, the endless policing of what counts as a human, the defence of *Homo sapiens* from the Neanderthals.[33] Racism, to name but one instance of prejudice, is when you try to pretend that there is a clean difference between you as a human and other, friendly beings 'over there' on the other peak, the one we call Nature. Because that means you have created a steep and profound valley in which all kinds of related beings are trapped so that you can't see them. If you like, you can have subjects (us) versus objects (Hitler's dog, R2D2, those face-less industrial robots, stones) because you have *abjects* (the beings in the valley) and you have 'disappeared' them. You look across the valley at R2D2 and see that he's very different from you (speciesism) because you have hidden all the uncanny intervening beings in the shallow valley between you and the cute little robot.

As the valley flattens into the plain, everything gains back a little bit of the abjection you were trying to dispose of down the toilet of the valley.

There are some basic rules of politeness on the Spectral Plain, and these have to do with the idea of *hospitality to strangers*. On what does such a hospitality depend? Ultimately, it depends on the weird idea of being hospitable to some being you couldn't possibly be hospitable to. There is a sort of impossible, spectral hospitality to the inhospitable that haunts the more straightforward kinds of hospitality, without which it would be sunk.

The deep reason for the necessarily veering quality of attunement, its oblique, slipping and sliding style, is that the beings to which it attunes are themselves slippery and uncanny. Evolution presents us with a continuum: humans and fish are related, so that if you go back far enough, you'll find that one of your very, very distant grandmothers was a fish. Yet you are not a fish. Wherever we slice the continuum, we will find paradoxes like that. Lifeforms are irreducibly uncanny – this means that the more we know about them, the *stranger* they become; science doesn't make it better, science makes it worse. This is why I coined the term *strange stranger* to refer to them. We find ourselves in the position of host, permanently. And hosting depends on an uneasy sense of welcome – who's going to show up through the door? The word *host* stems from a Latin word that can mean both *friend* and *enemy*.[34] We literally host all kinds of beings that can flip from friend to enemy in a moment – that's what having an allergic reaction is all about. Symbiosis, which is how lifeforms interconnect, is made up of all kinds of uneasy relationships, where beings aren't in total lockstep with one another.

X-Ecology

There is a sort of ethical and political Uncanny Valley too. What happens when we let the spectres out of that Valley, the spectres that haunt us with supposedly divergent versions of what counts as human? What happens when it becomes an ethical-political Spectral Plain?

When care is ramped up, stripped down, simplified in order to boost its energy – so we think – it loses some very precious qualities. Let's think again about that CARE/LESS calligraphy I was describing towards the start of this book, a beautiful encapsulation of the issue, in which seeming 'careless' might blend into being 'carefree', and where some modes of 'care' might end up being too heavy-handed. I'm not saying you can save Earth by playing videogames on the couch. I'm saying that being ecological, which is what this book is all about, isn't the same as being religious in a tight way, even though it isn't the same as being an atheist in a tight way either – because that's just upside-down religion. Since organized religion is an agricultural-age way for agricultural society to understand itself, it is riddled with the kinds of bug that have helped to destroy Earth. 'Store up your treasure in heaven' (as Jesus advises) means you don't need to worry so much about what happens down here, because it's less real and less important. Heidegger observed that Christianity was Platonism for the masses. I'm observing that, historically speaking at least, Platonism is Neolithic theism for the educated elite.

It's the same as how truthiness haunts truth. You could imagine this ambiguous care/less care/free quality as a spectre, like the spectres on the Spectral Plain, a sort of ethical spectre. It weirdly shadows and doubles and undermines and reinforces it. In short, it's a bit of a problem: but trying to shave this penumbra off and achieve a more smooth-looking form of care creates bigger problems. The care/lessness of indifference haunts care. But if we exorcise that ghost, we're back to survival for the sake of survival, and how's that been working out so far for life on this planet? We are so busy, and our current neoliberal machinations are just the latest upgrade to a busy, busy mentality that has been gripping us since 10, 000 BCE. The one emotion we love to hate in the media is apathy.

I recall, as a proud (?) member of Generation X, how we

were being told we didn't care enough about anything through the 1990s. It's funny, because as I looked around as the twenty-something me, I saw a lot of care in the 'civilized' world: people getting depressed by modern working conditions, people going into despair about environmental issues, nuclear families going subatomic, teenage years now extending to the age of thirty. Against the happy-happy enforcement of care, seeming a bit slack (a term we now use as Richard Linklater's film *Slacker* uses it) was a wonderfully refreshing stance.[35] I guess we could distinguish between claustrophobic, plastic forms of care, and more aerated, flexible ones.

I love being an X-er. The advertising, PR-type people who come up with these labels didn't know what label to slap on us, because we weren't behaving as we should. It's interesting if you are in the lineages of deconstructive philosophy as I am (Heidegger, Derrida and on). When Heidegger writes the word *Being* he puts it under a letter X, a gesture that Derrida calls *putting under erasure*. You can't say *Being* positively with a straight face, it makes Being look all bloated and solid like a huge blank bar of bland soap.

The CARE/LESS is the halo of care, its aura. When it gets hand-wringing, ecological talk retains a strong smell of the agricultural-age machinating that got us into this mess – it's that huge blank bar of bland soap again. I don't want to live in the world that kind of machinating would bring about. It would make the ways in which this current world sucks (to use a Gen-X term) look like the best thing that ever happened to anyone. I'm talking about a world based on greater and greater efficiency, greater and greater control of energy.

You can see this is how some people think about an ecological society. Instead, I think it's a world in which we can be so much more generous and creative than we've ever been, so much less 'caring' in that way that is hostile to actual lifeforms: survival mode.

Plastic care, stripped down and efficient, is highly toxic, especially when you scale it up to Earth magnitude and operate like that for 12,500 years. What is required instead is *playful care*. This doesn't mean care that is cynical. We actually have quite a lot of that: big corporations now enforce 'fun' in a most coercive manner. You are supposed to sing company songs or participate in collective team-building activities, or use videogame-like interfaces for working ('gamification'). We need something like the inverse, something like a *playful seriousness*. This mode would have a slight smile on its face, knowing that all solutions are flawed in some way. Expanded care, care with the care/less halo, is more likely to include more lifeforms under its umbrella, because it is less focused on sheer survival. The contrast we sometimes draw between selfishness and altruism is made from within a streamlined care outlook. You think there is a self and that therefore it needs protecting and boosting, and that caring for things that aren't the self would therefore involve some almost impossible to imagine emptying of the self, which in some agricultural-age religious domains is called *kenosis*, the Greek for 'emptying'. That doesn't sound fun and it doesn't even sound possible. It's a set-up. It's like how people are scornful about Buddhism – how can you desire to get rid of desire?

If I don't get behind this expanded care idea, then really,

this whole book has been a big waste of time. Because while I've been letting myself off the hook and not yelling factoids at you, secretly I'm not letting *you* off the hook and secretly I'm preaching to you, trying to convert you in a sneakier way. I'm machinating, but under the radar. That would mean that the whole way I wrote this was actually the opposite of playful seriousness: it was serious playfulness, goal-directed and 'fun'. I'd be trying to persuade you, and I think believing means holding on for dear life, and this is just a sales pitch.

So, in fact, I meant it all along, dear reader. I meant it when I said you didn't need to delete your indifference. You are quite right. You work so hard and you get so little in return, you have to smile relentlessly at work, you have to be your own paparazzo and upload a selfie to Facebook every five minutes, you have to 'Like' (that button) the right sorts of thing. In Freudian terms, your poor little ego is under attack from both sides, from the impulses of the id and the demands of the superego, both irrational and often superimposed, in our culture of 'repressive desublimation'.[36] And now I'm asking you to get all frantic about polar bears too? On top of everything else? So much frantic clicking, so much preening of exactly the right thing to say, a goal whose posts change every day, like the statistics. The thing about the superego is, it's impossible to fulfil its demands. Is it a feature of our psyches or a bug? Whatever the case, it's been inflamed by agricultural-age religion and its current ecological incarnation is therefore, however well meaning, a way of perfuming ecology space with exactly the wrong smell: the smell of busy, busy, zealous, industrious, 'just keep swimming, just keep swimming' intensity.[37]

Perhaps some of us care in all the wrong ways – too aggressively, too melancholically, too violently. Heidegger argues that even indifference is a form of care.[38] Perhaps indifference itself is pointing to a way to care for humans and nonhumans in a less violent way – simply allowing them to exist, like pieces of paper in your hand, like a story you might appreciate – or not – for no reason.[39]

I meant it, dear reader. Your indifference contains ecological chemicals, so don't throw the baby out with the bathwater. Actually, maybe you need to keep the baby *and* its ambiguous bathwater, and throw out the idea that you need to throw things out at all. In the final chapter, we'll be examining a few current styles of throwing-out in the name of being ecological, and we'll be contrasting these with being ecological in a way that doesn't reject ambiguity.

CHAPTER 4

A Brief History of Ecological Thought

It's a normal school day, and Homer Simpson is driving his son Bart and daughter Lisa with their friends to school. The radio is playing. Homer recognizes the music from his youth, and the kids want him to change the station. Instead Homer continues to embarrass them by launching into an account of the history of seventies rock bands. One paved the way for another, culminating in The Alan Parsons Project, 'which I believe was some sort of hovercraft'. He gets lost in the pedantry of how one band evolves into another one. He teaches the kids how to enjoy the music in ways that make no sense to them. He thinks he's being cool; they think he's being excruciating.[1]

And that is how most chapters like this go, with less humour and more seeming authority, to the point where you just can't stand them. It's what happened to me when I was asked to write a book like this – I immediately conjured up a picture of Homer explaining Grand Funk Railroad to his bemused and embarrassed kids. The book equivalent would be something with call-out boxes and 'easy-to-understand' categories that fit chapter headings. Actually, in the end, these sorts of thing are rather infuriatingly complicated, because they're not thought through.

Imagine you are in a record store – assuming they still exist; even better, imagine you are on iTunes or Spotify or some other online purveyor of music. There is a bewildering number of genres for you to choose, and the concept of genre is itself bewildering. Consider only one relatively narrow generic range. What on earth distinguishes electronica from electronic, techno from EDM? What does the iTunes category '90s music' actually mean? If it means 'any music recorded in the 1990s', it doesn't mean very much. What happens to music published in December 1989, or January 2000? What happens to music written in 2010 that derives from or alludes to music written in 1995? Is that '90s music'? Why not?

There are many ways to make this chapter tedious and inaccurate. First among these is what I shall be calling *the record store approach*. The record store approach is plagued with unexamined philosophical assumptions. It's difficult to read history sometimes, because it's always informed by implicit concepts that are often left unexplored. That's the main reason why we're not going to be using it in this chapter. Familiar demarcations are all too familiar. Sometimes we need to shake them up with thinking.

The record store approach consists of a bunch of preformatted labels that we simply employ without examination. What we would be dealing with in that case would be someone else's (or quite possibly a group of someones') way of thinking about ecological thought, without the merit of examining that someone. We simply inherit her or his categories without question. Then the categories get circulated, and become more legitimate. Then it becomes difficult to think outside the boxes of the categories we are retweeting.

Which in turn means that there are well-worn pitfalls and wells in the thought terrain – all kinds of fake paradoxes and problems, for example. Consider the clichéd discussion of 'nature versus nurture' that often takes place in popular media. It gets in the way far more than it helps.

I shall be organizing this chapter in a unique way. This mostly means that I shall not be organizing it according to the record store approach. Instead, let's return to and take seriously the *horizon* part of the idea that a genre is a *horizon of expectation*. Being bounded by a horizon implies that you are located somewhere. You are coordinating that line of trees, that mountainous ridge, those clouds with your body, your position. Being in a horizon like that implies having a certain stance, which is a metaphor for having a certain *attitude*. This seems like a much more precise, and also more toothsome, way of proceeding. Ideas come bundled with attitudes, remember. So rather than narrating a story, we will be exploring different *styles* of thinking, different ways of holding ideas. The beauty of the approach is that this way, we can allow for what happens in real life, namely that people hold a variety of overlapping and contradicting attitudes.

This is the reason why, in this chapter, we're going to be ignoring the self-labelling. We are going to be ignoring the sacred cows. Don't take their word for it. Otherwise you say the same thing over and over and the same guys get name-checked. What we have in that case is undigested history making its way through some preformatted digestive system, a history of ideas or worldviews or what have you. And this, however many footnotes it contains, would be just like Homer Simpson's account of Jefferson Starship. No grand tours, then, just like no information dumps, and in a way for the same reason. Those kinds of things are problematically unaware of the all-important *modes* in which they are telling us stuff, really important stuff.

What we're going to be doing instead, rather than looking at ideas 'in' time like marbles 'in' a prefabricated box, is looking at something like different *orientations*. We are going to study *styles* of being ecological in *thought mode*. One style thinks the world is going to end really soon. Another style thinks humans are unimportant. These orientations can overlap, because unlike worldviews they don't imply a shrink-wrapped, rigid system in which everything is a symptom of some explosive holistic whole that is greater than the sum of its parts.

Ways of being ecological summon certain kinds of words, certain kinds of arguments: in one philosophical view (Lacan, Althusser) they are called *subject positions*. In this case, far from being impressionistic or 'subjective', the phenomenological approach is more accurate: exploring the question, 'What is it like in heavy metal world?' might give you a lot more to chew on about heavy metal than an exhaustive account of all the types of metal according to the lingo that's evolved (black, death, speed, doom, grind . . .).

And thus it has come to pass that this chapter contains no details about ecological ideas. If you think about it, how people self-describe, especially if they are trying to fit their product into a record store, is *never accurate*. This is because of what we know about what phenomenology calls 'style' or what neurology now calls 'the adaptive unconscious'. You never see all of yourself all at once. This is how comedy works. Comedy is funny because the comic character can't see all of herself by definition. In trying not to be herself, she ends up manifesting herself, despite herself. So we're not going to have any 'deep' or 'shallow' or 'bright green' or 'eco-terrorist' or 'postcolonial' anything. Those are just record store labels.

But this chapter *does* contain ideas *about* ecological ideas. What do I mean? Let's have a look. While you look, realize that you can find these styles everywhere: in magazines, online, in what people say, in art, music and architecture, in patterns of behaviour and public policy . . . What I've done here is to isolate the active ingredients of each style, no matter where they manifest. We call this approach phenomenological reduction, which means exactly bracketing off everything except the colour, flavour and momentum (metaphorically speaking) of style as such.

The Immersive Style

Take, for example, the very basic idea of *being in environment at all*. Perhaps it would surprise you that this idea has a lineage and inculcates certain ways of thinking and feeling, ways that are not necessarily that great for actually existing lifeforms.

Would it surprise you to learn, for example, that this idea can be traced back directly to the earlier days of agricultural society? Doesn't that make perfect sense? There you are, settled in a city. Your ancestors were hunters and gatherers and nomads, but some time ago your more recent lineage joined the rest of the crew and settled down. You are looking out at things from your house. You are surrounded by things. You imagine these surroundings as something that swirls around your house every year, a sort of dynamic circle. You call it the *periechon*, which means literally *the thing that is going around.*[2]

There are lots of words that determine what we think an *environment* is to the feeling of being settled in a city. But in fact the environment is environ-ing, it's veering around. Another dynamic swirl.

Take the word *ambience. Ambo* is Latin for *on both sides* and again, the *-ence* suffix alerts us to something dynamic, something with a certain style. Ambience is the thing happening on either side of us, which makes sense if you're living in a house. The very word *ecology* comes from the Greek *oikos*, which means house, so in a sense people think *ecology* means *the rules of the house* or *how the house works* or *the truth about the house* or something like that. It's a funny house though: the walls are thick and spongy and retain all kinds of things we might not want in there; the roof is perforated, and other houses seem to be overlapping with ours. In a way, the house image, and the image of something circling around us (as long as it's not veering), is exactly the *wrong* way of imagining ecological coexistence. (As I hope this book has been demonstrating.)

One thing this type of ecological thinking seems to want to do is convey or express or explore some sense of *immersion* in something-or-other. What this something-or-other is precisely has changed over the years, but the basic style has coordinates that we can map.

When you think about DNA expression, what effects

genes produce in the world, you start to realize that it doesn't stop at the tip of a lifeform, but continues some way out from there. For instance, a spider's DNA expression (the spider's *phenotype*) doesn't stop at the tips of its legs: the spider phenotype stops (at least) at the tip of the spider's web. Spiders build webs because spider genes enable web-building. So a spider's genes don't just determine the shape of its body. A beaver's phenotype goes all the way up to the edge of its dam.[3] The human phenotype seems at present to cover a large swathe of Earth's surface, down quite a way into its crust as well at this point, which is why we call our current geological era the Anthropocene.

So when we think about the environment now, something interesting happens. When you look for the environment above and beyond lifeforms, you don't find it. Even the rocks and even the air you are breathing are part of some lifeform's phenotype. You are breathing because of an environmental catastrophe called *oxygen*. The Oxygenation Catastrophe occurred because oxygen is bacterial excrement, if you like – it's an unintended consequence of their success that anaerobic bacteria actually made their own environment poisonous long, long before humans did the same. (That's not the same thing as saying that humans *should* destroy their environment *because* they're successful or that destruction is inevitable.) So they eventually evolved to hide in other single-celled organisms, and these became the mitochondria, the animal energy cells, and the chloroplasts, the plant energy cells (and are the reason why plants are green). That's interesting in itself, isn't it? In a way, the fact that you are breathing is also a bacterial phenotype. And how green everything looks, in our idyllic picture of Edenic ecological utopia, is a bacterial phenotype. It's so amazing that you might accidentally hit your head on an iron railing while thinking about it, and since iron is another bacterial phenotype, you still wouldn't be free of our friends and enemies, our hosts and parasites, the bacteria.

The Style of Authenticity

Then we have countless ways of *writing* about ecology, by which we might loosely mean representing or otherwise exploring in sound, with paint, with words, and so on. No prizes for guessing what the favourite mode has been in the USA: it's the first-person narrative. There's a whole style of ecological thinking that goes along with this genre, and it's worth exploring, mainly to figure out how to avoid it – and why. You find it perhaps most vividly in what has sometimes been called 'nature writing', a quintessentially American ecological style. Others (of course) also employ it, but the Puritan resonances of an uncorrupted, providential 'wilderness' are definitely sourced in that country's first white settlers.

I'm going to call this one *the style of authenticity*. That's because, according to this style, the most important thing is to be genuinely authentically ecological – and so you need to say that you are, first to yourself and then to others. This style is associated with representation: it's about *authoring* yourself, so it's often about being an *author* (of writing).

Now the wonderful thing about first-person narratives is (take it from someone with a literature PhD) that they are *intrinsically unreliable*. 'Intrinsically' means structurally, which means (my paraphrase) 'no matter what you think about it and no matter how the author tries to wiggle out of it'. There is never a way to prove that the I who is doing the narrating totally overlaps the I that is being narrated. This basic feature of the first-person narrative comes in very handy all the time, otherwise you'd be stuck being exactly what you just said you were and exactly how people view what you say about yourself. You and your selfie would be exactly the same, and that would not be great. If reality coincided with its image, nothing could happen. Luckily you can say, 'I am bored', then you can say 'I am interested'. At an extreme you can even say, 'I am lying', and because of the irreducible separation between the speaking I and the spoken I, you will not implode.

Something funny also happens because of this, when you try to authenticate your first-person narrative. You think that if you add more and more details, people will believe you. But the more details you add, the stranger your description becomes, or the more desperate you look, and your tactic fails. It also fails in the case of nature writing because as you try to describe an authentic nature (as well as an authentic you, double trouble) you end up with more and more and more *words*, your shtick is that you're the kind of person, you insist, who doesn't like sitting around in some darkened room with a laptop; no, you're the kind of person who likes to be out there, roughing it in the desert or wherever. So you resort to some kind of journal style with date stamps, whether they are highly detailed and explicit or just implicit in the time sequence.

The speaking I and the spoken I are structurally different. You can't collapse the one into the other – well, you can, but this involves something called *Romantic irony*, which I will describe in a moment. But this is just the kind of feature, so wonderful and so essential to *enjoying* a good memoir or narrated movie, which is exactly what environmentalist prose tries – and repeatedly fails, because it's inherent to the first-person form – to edit out. It's a bit like trying to saw yourself off the branch you are sitting on; literary richness sitting on exactly this branch. It makes no literary sense, and actually it makes no ecological sense, because an artificially flattened, trying-to-be-sincere (and therefore being unintentionally funny) first-person eco-narrative (think of the earnest nature-writing journal or travelogue) is how you make the world into your candy bar or packet of tortilla chips, and everyone gets to watch you sit on your couch (which you pretend is called *wilderness*) and eat them.

The poets of the British Romantic period knew how dodgy the first person was, which is exactly why they used it. It's just not correct to think of them as naive nature writers, as we too often do, even though they did tell stories about encountering mountains or hearing the terrifying yet invigorating sound of the surf. In fact, they were trying to get past all that pretty nature stuff, which was old by the time they started. The age before the Romantics was called the Age of Sensibility, a moment at which European scientists discovered the nervous system and developed all kinds of theories about how meaning arose in an unmediated way from the sensations. Nature meant something you feel spontaneously, something that doesn't require any hesitation or reflection to grasp, something that underlies the necessarily false artifice of society and what the Age of Sensibility often called 'custom'. Consider Rousseau, for example, who argues that humans are naturally free, but society causes them to be enchained.

The slippery nature of the first-person narrative is exactly what these poets and prose writers fully folded into their work, with the narrators sometimes even alerting their readers to the fact that they had been lying to them, or luring them in and then proving that they were not to be trusted. Romanticism doesn't mean having your head in the clouds: this approach suggests a less anthropocentric attitude, and one that was in fact more in line with scientific curiosity; they were laying bare how their stance changed what they were seeing. Think of the difference between viewing a cliff from afar, seeing it as a distant object that incites a sense of awe, compared with getting up-close to a rock face with a magnifying glass, looking at it in detail and deconstructing its mighty mass. The eighteenth-century equivalent of the iPhone camera and the selfie stick – in the sense that people would take it around when they travelled – was called the Claude Glass. The Claude Glass was a hemisphere of sepia-coloured glass; you got into a special position for looking at the landscape in a prescribed way, and then you looked into the glass. Upside-down, you saw reflected the landscape you were beholding, as if painted in sepia ink. Unlike the Claude Glass the magnifying glass makes the rock face look very strange, because it is no longer fulfilling our anthropocentric requirements as a nice background (totally unlike our selfie).

Similarly, the Romantic poet gets up-close and personal with her or his own experience, which in a way is the inner equivalent of the rock face. Experience never has a 'This is a [insert your name here] experience' running through it like a bar code or a copyright mark or one of those phrases you see in a stick of rock at the British seaside. It lacks this bar code *especially* when the experience is really intimate. Imagine being in a car crash. It's so vivid: it's a trauma. Precisely because of this, there is a feeling of *unreality*. The feeling of unreality goes hand in hand with the less you-scaled, more ego-shattering event, which becomes part of you (scars you for life, perhaps) and one of your most vivid, even treasured (perhaps in a bad way) memories.

So the gyrations of nature writers can be massive regressions from a style we should have all learned from by now – the powerful ambiguities of a William Wordsworth, the haunting multiple voices of a Charlotte Turner Smith, the weirdly ecological ennui of a Charles Baudelaire. And for exactly the same reasons they don't add up to being ecological, which is the supposed point. Because being ecological includes a sense of my weird inclusion in what I'm experiencing; it isn't an unmediated, direct experience at all.

The Religious Style

If even the *concept* of environment is a Neolithic product and thus part of the problem and not part of the solution, perhaps we should spend our ecological time bemoaning the horror of so-called civilization? A certain style arises that I am going to call *the religious style*. This mode becomes more and more popular every day and its modus operandi becomes increasingly rabid. Social media, for instance, has become a place of ever-increasing judgemental differentiation.

The religious style has a long heritage. Consider, for example, that popular literary genre, the pastoral. In this genre, a couple of shepherd-type people – they have a nomadic quality to them, so perhaps this is why they are used – go up a hill and look down on the awful corrupt things happening in the town below, lamenting the general badness of civilization. Usually the ecological way of being religious takes the form of some kind of misanthropy, which is still anthropocentrism: humans are evil because they have caused ecological destruction. This idea is hardwired into accounts of what Judeo-Christian religions call the Fall, but also other agricultural-age religious accounts of the move towards agricultural society, such as Hinduism. In a way, perhaps all anthropocentrism is misanthropic, because it ultimately does a disservice to humans too. Perhaps we should be calling it *misanthropocentrism*.

Hegel has a vivid way of describing this religious style: he calls it *the beautiful soul*.[4] For Hegel, knowing comes in all kinds of flavours, and this means that ideas and their flavours are always a bit out of balance, like a slinky perched on a step. How-to-think the idea and the ideas as such are necessarily different. This imbalance causes the idea-plus-flavour manifold to flop over itself, like said slinky. The basic imbalance that characterizes the style of the beautiful soul is something like an ultra-religious person, someone you might call 'religiose'. This kind of person sees the world as evil, or, better, regards evil as a thing that she or he can get rid of. Evil isn't part of me, it's something lodged in me that I can dispose of. Can you see what the imbalance is? The style is out of balance because the gaze that sees evil as a thing 'over yonder' *is exactly evil as such*. Think about how al Qaeda saw America as the source of all evil on Earth, and conversely about how the US Bush Administration saw the same in al Qaeda. When you see evil as a thing apart from yourself 'over there', you can fly a plane into it or destroy it with a powerful bomb. You can justify murder. Evil is the gaze that sees evil as a thing apart from me.

This is a typical and bad side effect of all kinds of environmentalist viewpoints. Think about the view from the edge of the Solar System that Carl Sagan called the 'pale blue dot': a picture of Earth reduced to just one pixel. It's the last photograph of Earth taken by the *Voyager* space probe as it left the Solar System in 1990.[5] Sagan does what some Enlightenment writers did, framing human events as tiny, petty things that take place against this vast, indifferent backdrop: the point being, we shouldn't be so concerned with our human-centred business, we should be more peaceful and loving, and so on. But the attitude within which this supposedly hippy style is staged is precisely that of the evil gaze – isolating everything bad into a tiny dot, a single pixel in the gigantic picture of the universe, a position of infinite contempt and hostile judgement.

The truly spiritual position is to realize that *whatever evil is, it is an intrinsic aspect of oneself*. This is equivalent to noticing that we are made of and surrounded and penetrated by all kinds of beings which, in the right combinations, might do us a great deal of harm. In other words, it's equivalent to the uneasy hosting we discovered to be the essence of symbiosis. And this entails that many forms of environmentalism aren't really very ecological at all. They try to find fault by isolating one particular entity – say a large corporation that makes toxic products, or a particular kind of consumer, or consumerism as such – without considering how the entity is caught in all kinds of networks and systems. Who is to blame for global warming, the Americans who invented air conditioning, or the Chinese and Indians eager to use it? This isn't to say that some beings are not more to blame than others. Humans caused global warming, not sea turtles. It's *how* we think this blame that is key.

The Efficient Style

Alternatively, you might not be concerned with good and evil, at least not directly. You might see the ecological realm as a domain that needs to be well maintained: your ethical or political spectrum runs from efficiency to inefficiency. Your approach is normative, like the religious style, but not as explicitly: you value a smoothly functioning biosphere optimized for human existence without too much damage to other lifeforms.

This is going to be the longest account of ecological styles in this chapter. This is both because it's a very popular style, and because it's got a lot of moving parts.

You don't have to be keen on geoengineering to perform this style. Geoengineering, which since about 2000 has become a popular way of imagining how to solve the biggest ecological problems, means interfering with the biosphere at the largest possible, planetary scale. For example, technocrats might decide that the best solution to global warming is to put gigantic mirrors in space to reflect back the Sun's heat, or to fill the ocean with iron filings to encourage the growth of phytoplankton such as photosynthesizing algae. The seduction of this approach is the sense of mastery it bestows. The trouble is that since any geoengineering action affects the whole of the biosphere, there can be no reverse gear. There is no way to check in advance exactly what will happen, and there is no way to undo it once it has happened, if by 'undo' we mean 'completely erase its effects'.

Geoengineering is just one way in which someone might perform this ecological style. It's instructive, because it brings to light the dominant way in which Western philosophy has imagined how reality works for the last two centuries, a form of thought named correlationism, which we've explored somewhat. Correlationism, the idea that the world isn't real until some correlator (usually tied to a human being in some way) has 'realized' it, can produce the fantasy that reality is a blank slate waiting for (human) projections to fill it in, like a movie screen waiting for a movie to be shown on it. The idea that the world is a blank canvas waiting for the correlator to paint on it is rather obviously ecologically violent: the world is not a blank screen, it's a coral reef, it's a high-altitude Alpine ecosystem, it's a humpback whale.

A less extreme version is the idea that it could be dangerous to imprint the world with (human) desire, as if it were a blank slate and as if we knew what was good for it. This version tries to minimize the impact of the correlator, to tread lightly, to be efficient, to minimize one's carbon footprint.

While admirable, and in many respects quite right, this style has its limitations. It's a very popular way of being ecological. It's attractive, because it's based on the idea of attunement that we explored in the previous chapter. Like a boat floating with the movements of the ocean, this style of efficiency tries to minimize energy use by tacking close to what is already the case, like steering a ship without exerting too much effort. This style of efficiency is a dynamic dance that attends to how the momentum of the world is at any particular moment, and is inherently on the side of the status quo. It is prevalent in theories of social systems based on cybernetics – the Greek word *kubernētēs*, like the word 'governor', from which cybernetics comes, means pilot or steersman.

Governing or mastering through tracking, tacking, adhering to . . . such concepts also evoke fantasies of mastery. The idea is that one could 'get it right'. But if the system is dynamic, temporal, getting it right never stays still. The idea is close to the more open concept of attunement, which is like what happens when one is playing music with others: you figure out that music is first and foremost a kind of *listening*. But the difference is that the efficiency approach must always be based on some kind of pre-established parameters as to what counts as efficient. The idea is to eliminate mistakes, which boils down to the elimination of the difference between the pre-established past and the open future. Efficiency stifles *creativity*, which is a more basic way of thinking about attunement. Thus might be born a certain kind of ecological 'lifestyle', a way of constructing a certain kind of world that appears to function smoothly, based on a fantasy that something close to perfectly smooth functioning could ever be achieved.

But this smoothness is only smooth from the point of view of a particular scale. My smoothness as I manage to park my car ever so nicely in a narrow spot is a horrible *malfunction* from the point of view of the snail whose shell my car wheel is crushing. The idea that there are multiple worlds because there are multiple lifeforms and that no one world or scale is the 'right' one means that efficiency is only efficient from a particular standpoint. For example, the idea of sustainability implies that the system we now have is worth sustaining. It implies furthermore that 'continuing for a longer time' is a hallmark of success, which in turn implies a model of existing having to do with *persisting*, going on, being constantly present. But we've established that things aren't like that. So in the end the style of efficiency is going to be stifling and uncreative, not allowing for malfunctions and accidents, which are ironically much more like the way things actually are. It's not the case that things are just functioning smoothly until they don't. Smooth functioning is always a myth.

Bataille gave a name to this smooth functioning myth: the restricted economy. A restricted economy is one in which the dominant theme is efficiency: minimum energy throughput. The Earth is finite, and economic flows must be restricted to its finite size and capacities. So much ecological ethics, politics and aesthetics is based on the economy of restriction.

Although it sounds very reasonable, something is drastically missing from the style of restricted economy, which means that in the end it's at the very least spiritually unsatisfying for those who try to maintain it. Because malfunctioning is deeper than (smooth) functioning, there is an excessive intensity to the energy of things that just can't be contained efficiently. There is a lack of attention to *what* is being efficiently sustained. And as the model of efficiency will always be a little bit behind the times (if only by a few moments – you can't be radically proactive because you need to gather data about the current situation in order to work with it in an efficient way), it won't ever accurately track the way things are, despite the promise that it could.

Artists of all kinds and practitioners of esoteric spiritual traditions intuit this problem. In those traditions, the aim is not so much to get rid of or even to transform negative emotions, but to embrace them and discover the energies within them that transcend the ego: the ego is taken to be the big problem, not the perceptions or the kinds of phenomena that are arising, such as anger. Again, it's not what you're thinking, it's *how you're thinking* that causes suffering. Anger can happen, and if you don't cling to it, it becomes just another colour or flavour of energy. This isn't about pushing away or denying one's emotion, but rather about exploring it without too much clinging. If you do cling to it, it feels just awful; it's 'my' anger, how am I going to get rid of it . . . Something like this insight needs to be part of being ecological, otherwise the risk is that humans will create a control society (to use the technical term from Deleuze) so intense that, as I said earlier, the current one, already very hard to bear, will seem loosey-goosey by comparison.

Moreover, the ultimate horizon of efficiency is *petroculture*: the fact that oil, a precious toxic resource, dictates how we conduct ourselves. In a world without oil, we shouldn't be imagining ecological action in the key of oil. That would be behaving according to an energy economy that no longer exists. And this wouldn't be fun at all. I think that ecological politics is about expanding, modifying and developing new forms of *pleasure*, not restraining the meagre pleasures we already experience because we are only thinking in ways that our current modes of doing things allows. What would pleasure look like beyond the oil economy?

Last year I switched my house's energy plan from one that relied on fossil fuels to one that relied only on wind (Texas has a surprisingly vast array of wind farms). For the first three days of being on this new plan, I felt incredibly smug and virtuous. I felt pure and efficient. I felt as if I understood finally what *sustainability* meant. And then . . .

I realized that I could have a pumping disco in every single room of my house, and far, far fewer lifeforms would be harmed at all, compared with burning fossil fuels to power just the basics of my house. Solar and wind power would mean no carbon emissions, which would mean less or no global warming (depending on how many people used solar power), which would mean less or no extinction of lifeforms. And being dead is a terrible inconvenience if pleasure is your goal; just keeping lifeforms alive is allowing for some kind of pleasure (and don't forget the pleasure of enjoying their existence, and the pleasure of doing less harm). And then I realized that this sort of feeling would be what living in an ecologically attuned society actually feels like. Instead of policing pleasure we would be inventing new ones.

This means something almost unbelievable. (The question of why it's unbelievable is itself interesting. We'll get to it.) Brace for impact.

An awful lot of ecological speech is actually oil economy speech. In fact, almost *all* ecological speech isn't ecological speech at all. Ecological speech is deeply distorted by the oil economy we live in. All that language about efficiency and sustainability is about competing for scarce highly toxic resources.

But if you think modern life is tight and restrictive and full of all kinds of police and policing, hold on to your hat. Imagine what an ecological society based on those principles of restriction and efficiency would feel like. I would like very much not to live on Earth if that is the direction in which we go.

Working with Paranoia

Ecological awareness presents us with a disturbing fact. In ecological awareness, 'away' has disappeared, because we know, for example, that our toilet waste doesn't go to some special different place called 'away', it just goes somewhere else. If there is no *away*, then there is no *here*. We have lost reality. The *-ity* part of that word is the most important part. And we can see exactly why, in the light of ecological awareness. It's not that there is nothing at all. We have the *real*. But it doesn't make any sense any more. That's the trouble with data dump mode, and it's the explanation for it. Data dump mode is just enhancing the incapacity of things to mean anything any more to us. Our awareness is no longer human-scaled, no longer keyed to anthropocentrism. This is potentially great, if we can 'own' and explore it. But it will require all kinds of trauma work to go through. It would be like trying to figure out how to exist now that we have become totally paranoid. This might be very tricky, but not impossible – people recover from trauma all the time. We would need to learn to become playful about the lack of an obvious solid ground of meaning, one obvious scale on which to see and act. Again, this is hard, but not impossible. In wartime people learn how to handle their situation, as difficult as it is. You can learn how to navigate through a bad dream. It means stepping outside of our comfort zone, but then again, some of our human comfort zones have been extreme discomfort zones for other lifeforms, and in the long run for us.

The End of the World

So, double trouble. Sure, we can fix the planet. But why? Psychically it's as if we are being crushed. And the modes we have to draw on that might restart things are part of the problem. Currently our ways of restarting reality tend to be based on severing our connections with nonhuman beings in every respect: social, psychic and philosophical. So we have inadequate political, technical and psychic tools at our disposal with which to fix things. But curling up in the foetal position in despair isn't going to work either. Instead of imagining that everything is useless and that the apocalypse has come – so there's no point anyway – and instead of thinking that we have to completely reimagine how to do things (we'll never get going with those attitudes), it would be better to start where we are and use some of the inadequate and broken tools we have, and see how they get modified by working at scales and with lifeforms that are unfamiliar to us, for which the tools were not designed. In the process, the tools might undergo some changes.

I am very against the fatalism of thinking that this is the end of the world, or that the end of the world is imminent. In a funny way, it's as if the end of the world has already happened, if by *world* we mean a stable set of reference points that guide our actions. Like Nietzsche proclaiming that God is dead, maybe we should boldly proclaim that the world is dead. Now that there is a bewildering variety of scales on which to think and act – ecosystem scale, planet scale, biosphere scale, human scale, blue whale scale . . . – it's already the end of the 'world'. This is actually a relief. It means we don't have to hold on to a fantasy for dear life, the fantasy of anthropocentrism, which is inaccurate and violent. It's like those horror movies in which the hero finds out that she or he is already dead. If you're already dead, there's nothing to be afraid of, is there?

Inconclusive in Conclusion

Being ecological is like being a teacher. When you first start teaching, you try so hard to teach that it becomes excruciating. You want your students to like you. You want to like them. You don't want to feel this excruciating feeling that you yourself are generating by trying so hard. You start to work with aggression (or you quit). You realize that you are a channel for your and your students' negative as well as positive feelings, and your job is to hold those feelings for the students' benefit. Then you wonder why you are trying so hard, and maybe you start to let go. You begin to trust. You begin to realize that you are a teacher, no matter what, because at least one other person knows you're their teacher. You can relax into that.

It's the same when you're a parent. You spend some time trying desperately to *be* a parent. And then once you realize that you just *are* a parent, you can relax. At least someone knows you're their parent.

You are a fully embodied being who has never been separated from other biological beings both inside and outside your body, not for one second. You are sensitively attuned to everything happening in your world, which is why you end up blocking some of it, because you are afraid the stimulation might be too intense. You have an idea that there is an inside and an outside of yourself, and perhaps this is the deepest way in which you start to think that being ecological involves some massive change.

Snared in the urgency of ecological awareness and the horror of extinction and global warming, it's so incredibly difficult to miss this key point. I can't tell you how many environmentalist conferences I've been to where the ending atmosphere had to do with some kind of fist-clenching, jaw-clenching desperation to be or do something totally different. What a set-up – once you've established this totally different space, you are already separated from it by a gigantic chasm, and being right or smart in this kind of world means showing yourself and everyone how deep and wide this chasm is. You've just made sure that you are *never* going to be ecological. The one thing that could help gets drowned out by the fear of the intensity of our reactions to the data input (oceans acidifying! Climate warming! Species going extinct!).

But you are already a symbiotic being entangled with other symbiotic beings. The problem with ecological awareness and action isn't that it's horribly difficult. It's that it's too easy. You are breathing air, your bacterial microbiome is humming away, evolution is silently unfolding in the background. Somewhere, a bird is singing and clouds pass overhead. You stop reading this book and look around you.

You don't have to *be* ecological. Because you *are* ecological.

Notes

INTRODUCTION: NOT ANOTHER INFORMATION DUMP

1. Sigmund Freud, *The Uncanny*, tr. David McLintock, intro. Hugh Haughton (London: Penguin, 2003).
2. Pink Floyd, 'Breathe', *The Dark Side of the Moon* (EMI, 1973).
3. Graham Harman's beautifully clear books include *Heidegger Explained: From Phenomenon to Thing* (Chicago: Open Court, 2007), which succinctly shows that Heidegger is actually quite easy to understand. What Heidegger is arguing is simply that *being is not presence.*
4. Derek Parfit, *Reasons and Persons* (Oxford: Oxford University Press, 1984), 355–7, 361.

CHAPTER 1: AND YOU MAY FIND YOURSELF LIVING IN AN AGE OF MASS EXTINCTION

1. W. D. Richter, dir., *The Adventures of Buckaroo Banzai across the Eighth Dimension* (20th Century Fox, 1984).
2. See Dipesh Chakrabarty, 'The Climate of History: Four Theses', *Critical Inquiry* 35 (Winter, 2009), 197–222.
3. Martin Heidegger, *Being and Time*, tr. Joan Stambaugh (Albany, NY: State University of New York Press, 1996), 59–80.
4. Heidegger, *Being and Time*, 17.
5. Talking Heads, 'Once in a Lifetime', *Remain in Light* (Sire Records, 1980).
6. John Keats, 'In Drear-Nighted December', in *The Complete Poems*, ed. John Barnard (London: Penguin, 1987), line 21.
7. Timothy Morton, *Dark Ecology: For a Logic of Future Coexistence*, (New York: Columbia University Press, 2016).

8. *Doctor Who*, 'Blink', dir. Hettie MacDonald, written by Steven Moffat (BBC, 2007).
9. John Cage, '2 Pages, 122 Words on Music and Dance', in *Silence: Lectures and Writings* (Middletown, CT: Wesleyan UP, 2011), 96–7 (96).
10. For a full discussion of this, see the end of Morton, *Dark Ecology*, 111–74.
11. John Carpenter, dir., *The Thing* (Universal Studios, 1982).
12. Davis Guggenheim, dir., *An Inconvenient Truth* (Paramount Classics, 2006).

CHAPTER 2: . . . AND THE LEG BONE'S CONNECTED TO THE TOXIC WASTE DUMP BONE

1. The Wachowskis, dirs., *The Matrix* (Warner Brothers, Village Roadshow Pictures, 1999).
2. Gregory Bateson, *Steps to an Ecology of Mind*, foreword Mary Catherine Bateson (Chicago: University of Chicago Press, 2000).
3. Edmund Husserl, 'Prolegomena to All Logic', *Logical Investigations*, tr. J. N. Findlay, ed. Dermot Moran (London: Routledge, 2008), 1.1–161.
4. John Cleese and Graham Chapman, 'Argument Clinic', *Monty Python's Flying Circus* (BBC, 1972).
5. In *The Ecological Thought* (Cambridge: Harvard University Press, 2010).
6. Georges Bataille, *The Accursed Share: An Essay on General Economy*, tr. Robert Hurley, vol. 1 (New York: Zone Books, 1988).
7. James Lovelock, *Gaia: A New Look at Life on Earth* (Oxford and New York: Oxford University Press, 1987).
8. The title of a famous sermon by Jonathan Edwards (1741).
9. William Blake, *Auguries of Innocence*: *The Complete Poetry and Prose of William Blake*, ed. David V. Erdman (New York: Doubleday, 1988), lines 1–3.
10. Ibid., lines 9–10.
11. Line from Talking Heads, 'Once in a Lifetime', *Remain in Light* (Sire Records, 1980).

CHAPTER 3: TUNING

1. The documentary *Crude* is worth watching to get a sense of what happened in Ecuador: Joe Berlinger, dir., *Crude* (Entendre Films, Radical Media, Red Envelope Entertainment, 2009).

2. Carl Safina, *Beyond Words: What Animals Think and Feel* (New York: Henry Holt, 2015), 29.
3. David Eagleman, 'Who Is in Control?', *The Brain*, episode 3 (PBS, 2015–).
4. The work of Jacques Derrida has been devoted to this theme in numerous ways.
5. George Lucas, dir., *Star Wars* (20th Century Fox, 1977).
6. One of the best places to start is Graham Priest, *In Contradiction: A Study of the Transconsistent*, 2nd edn (Oxford: Oxford University Press, 2006).
7. The phrase is the title of Lacan's twenty-first seminar.
8. Richard Dawkins, *The Extended Phenotype: The Long Reach of the Gene* (Oxford and New York: Oxford University Press, 1999).
9. Gillian Beer, 'Introduction', Charles Darwin, *On the Origin of Species*, ed. Gillian Beer (Oxford and New York: Oxford University Press, 1996), xxvii–xxviii.
10. The glass example is explored in detail in Timothy Morton, *Realist Magic: Objects, Ontology, Causality* (Ann Arbor: Open Humanities Press, 2013), 193.
11. Timothy Morton, 'Buddhaphobia: Nothingness and the Fear of Things', in Marcus Boon, Eric Cazdyn and Timothy Morton, *Nothing: Three Inquiries in Buddhism* (Chicago: University of Chicago Press, 2015), 185–266.
12. Colin Campbell, 'Understanding Traditional and Modern Patterns of Consumption in Eighteenth-Century England: A Character-Action Approach', in John Brewer and Roy Porter, eds., *Consumption and the World of Goods* (London and New York: Routledge, 1993), 40–57; Timothy Morton, *Dark Ecology: For a Logic of Future Coexistence* (New York: Columbia, 2016), 120–23.
13. Sigmund Freud, *Beyond the Pleasure Principle*, in *Beyond the Pleasure Principle and Other Writings*, tr. John Reddick, intro. Mark Edmundson (London: Penguin, 2003), 43–102.
14. Alex Proyas, dir., *Dark City* (New Line Cinema, 1998).
15. Ridley Scott, dir., *Blade Runner* (Warner Bros., 1982).
16. Georg Wilhelm Friedrich Hegel, *Outlines of the Philosophy of Right*, ed. Stephen Houlgate, tr. T. M. Knox (Oxford: Oxford University Press, 2008), 16
17. Jacques Derrida, 'Plato's Pharmacy', *Dissemination*, tr. Barbara Johnson (Chicago: University of Chicago Press, 1981), 61–171.
18. William Shakespeare, *Hamlet*, ed. T. J. B. Spencer, intro. Anne Barton (Harmondsworth: Penguin, 1980), 2.1.63 (p. 99).
19. Karl Marx, *Capital*, tr. Ben Fowkes, volume 1 (Harmondsworth: Penguin, 1990), 1.163.
20. Jeffrey Kripal, *Authors of the Impossible: The Paranormal and the Sacred* (Chicago: University of Chicago Press, 2010).

21. Aaron D. O' Connell et al., 'Quantum Ground State and Single Phonon Control of a Mechanical Ground Resonator', *Nature* 464 (17 March 2010), 697–703; Amir H. Safavi-Naeini, 'Observation of Quantum Motion of a Nanomechanical Resonator', *Physical Review Letters*, art. 033602 (17 January 2012).
22. The recent work of Anton Zeilinger has been devoted to eliminating loopholes in nonlocality theory, in other words, maintaining the paradox of two entities tuning to one another simultaneously.
23. *Oxford English Dictionary*, 'Weird', adj. http://www.oed.com, accessed 9 April 2014.
24. Immanuel Kant, *Critique of Pure Reason*, tr. Norman Kemp Smith (Boston and New York: St Martin's, Press, 1965), 84–85.
25. I am grateful to Tanya Busse for discussing this with me.
26. Alan Turing, 'Computing Machinery and Intelligence', in Margaret A. Boden, ed., *The Philosophy of Artificial Intelligence* (Oxford and New York: Oxford University Press, 1990), 40–66; René Descartes, *Meditations and Other Metaphysical Writings*, tr. and intro. Desmond M. Clarke (London: Penguin, 2000).
27. Brian O'Doherty, *Inside the White Cube: The Ideology of the Gallery Space*, intro. Thomas McKevilley (Santa Monica and San Francisco: Lapis Press, 1986).
28. O'Doherty, *White Cube*, 35–64.
29. The Beatles, 'Sgt. Pepper's Lonely Hearts Club Band', *Sgt. Pepper's Lonely Hearts Club Band* (Parlophone, 1967).
30. John Ruskin, 'The Seven Lamps of Architecture: The Lamp of Memory', in *Selected Writings*, ed. Dinah Birch (Oxford: Oxford University Press, 2009), 16–27 (25).
31. This is a deep implication of Hermann Minkowski's geometrical proof of relativity theory.
32. See, for example, O'Connell et al., 'Quantum Ground State and Single Phonon Control of a Mechanical Ground Resonator'; Safavi-Naeini, et al., 'Observation of Quantum Motion of a Nanomechanical Resonator'.
33. Giorgio Agamben argues about this in *The Open: Man and Animal*, tr. Kevin Attell (Stanford: Stanford University Press, 2004), 33–38.
34. There is a very profound essay by Jacques Derrida on this topic: 'Hostipitality', *Acts of Religion*, ed., tr. and intro. Gil Anidjar (London and New York: Routledge, 2002), 356–420.
35. Richard Linklater, dir., *Slacker* (Orion Classics, 1990).
36. Herbert Marcuse, *One-Dimensional Man: Studies in the Ideology of Advanced Industrial Society*, intro. Douglas Kellner (London: Routledge, 2002), 75–78.

37. I am quoting Dory from Andrew Stanton, dir., *Finding Nemo* (Buena Vista Pictures, 2003).
38. Heidegger, *Being and Time*, tr. Joan Stambaugh (Albany, NY: State University of New York Press, 1996), 40–41, 113–14, 115, 116, 127.
39. Some serious ecological philosophy points in this direction. See, for instance, Giorgio Agamben, *The Open: Man and Animal*, tr. Kevin Attell (Stanford: Stanford University Press, 2004).

CHAPTER 4: A BRIEF HISTORY OF ECOLOGICAL THOUGHT

1. Brent Forrester, 'Homerpalooza,' dir Wesley Archer, *The Simpsons*, Series 7, Episode 24 (Fox, 1996).
2. See Leo Spitzer, 'Milieu and Ambiance', in *Essays in Historical Semantics* (New York: Russell and Russell, 1968), 179–316.
3. See Richard Dawkins, *The Extended Phenotype: The Long Reach of the Gene* (Oxford and New York: Oxford University Press, 1999).
4. Georg Wilhelm Friedrich Hegel, *Hegel's Phenomenology of Spirit*, tr. A. V. Miller, analysis and foreword by J. N. Findlay (Oxford: Oxford University Press, 1977), 383–409.
5. Carl Sagan, *Pale Blue Dot: A Vision of the Human Future in Space* (New York: Random House, 1994).

Index

A

B

C

D

E

F

G

H